Marine Otter Conservation

Liliana Ayala
Raúl Sánchez-Scaglioni • Gonzalo Medina-Vogel
Editors

Marine Otter Conservation

Springer

Editors
Liliana Ayala
Carrera de Biología Marina
Universidad Científica del Sur
Lima, Peru

Asociación Peruana para la
Conservación de la Naturaleza
Lima, Peru

Gonzalo Medina-Vogel
Centro de Investigación para la
Sustentabilidad
Universidad Andres Bello Santiago
Santiago, Chile

IUCN Species Survival Commission
Otter Specialist Group
Gland, Switzerland

Raúl Sánchez-Scaglioni
Asociación Peruana para la
Conservación de la Naturaleza
Lima, Peru

ISBN 978-3-031-53930-5 ISBN 978-3-031-53931-2 (eBook)
https://doi.org/10.1007/978-3-031-53931-2

This Springer imprint is published by the registered company Springer Nature Switzerland AG
The registered company address is: Gewerbestrasse 11, 6330 Cham, Switzerland
Paper in this product is recyclable.

To researchers, managers, students, educators, fishers, and activists who have dedicated their lives to explore, understand, and conserve the marine environment. Especially to Jose Pizarro-Neyra, a Peruvian passionate biologist who studied marine otters and who initiated the history of this book.

Preface

Despite the enormous peculiarities of the marine otter, also known as the chungungo, this book notice how little we know about this unique species, which inhabits the coastal areas of Peru, Chile, and possible extinct from Argentina. This otter species is also known as the smallest marine mammal of the world and the most recently adapted to the marine life. The South American marine otter has a long latitudinal distribution, coupled with its adorable charisma and appeal, and has captured the attention of researchers, naturalists, and the public. However, like many other otter species, we not only have limited knowledge about its ecology, reproduction, and physiology, but also about its habitat needs and threats. The coastal marine environment where it lives in is inexorably altered, modified, and transformed to meet the demands of coastal development and the increasing human population, progressively endangering not only the marine otters but also the biodiversity that resides within.

In early 2022, José Pizarro-Neyra, a Peruvian biologist, naturalist, and educator actively involved in marine otter conservation, invited us to collaborate in a book that consolidates the level of knowledge about the species and provides elements that contribute to its conservation. José decided to entrust us with this task for personal reasons. The first thing that became clear in this endeavor was the limited number of researchers involved with this species, although their number in Chile is much higher than that in Peru. The second realization was the need to organize and synthesize information. Efforts to better understand the species and contribute to its management and conservation should be based on clear evidence, with ongoing measures requiring review and improvement.

Initially, we had to understand the environment in which marine otters live. This is described for Peru in Chap. 1, "Environmental Components of the Marine Otter Habitat of Peru," by Sonia Valle-Rubio and Aldo G. Indacochea, highlighting the abundance and productivity of the Humboldt Current System, as well as the physical and geological variability of the coast with beaches, rocky shores, and cliffs along the Peruvian coast. It is noteworthy that food availability is not limited, even with globally notable oceanographic events such as the El Niño Southern Oscillation.

A comprehensive description of the natural history is provided by Gonzalo Medina-Vogel, Carlos Calvo-Mac, Nicole Delgado-Parada, and Gabriela Molina-Maldonado in Chap. 2, "The Natural History of the Marine Otter (*Lontra felina*)." However, they emphasize the limited knowledge of the species. Leonardo Hostos-Olivera and Juan Valqui attempt to address some of this in Chap. 3, "Trophic Ecology of the Marine Otter (*Lontra felina*): Review, Quantitative Analyses and Population Implications," describing not only trophic ecology but also morphological variations in otter skulls due to changing prey along its extensive distribution. Medina-Vogel et al. also draw attention to the need to understand the metabolism and freshwater needs and limitation of this species, especially since it is found in areas where freshwater availability might be limited, as described by Joaquín A. Ugarte-Núñez and Ulrich Zanabria-Alarcón in Chap. 6, "The Marine Otter in Continental and Andean Habitats." How marine is the otter? Why was it not found in other rivers throughout its distribution?

In Chap. 4, "Marine Otter Conservation in Peru," Jeffrey C. Mangel, Joanna Alfaro-Shigueto, Clara Ortiz-Alvarez, and Carlos Calvo-Mac describe the conservation status of the species in Peru, while Chap. 8, "Marine Otter (*Lontra felina*, Molina 1782) Conservation in Chile," by Gonzalo Medina-Vogel, Carlos Calvo-Mac, Gabriela Molina-Maldonado, and Nicole Delgado-Parada does the same for Chile. Fortunately, the species has ceased to be hunted and fully protected, yet it remains endangered. Their irregular distribution, habitat fragmentation, and variation in population density along its distribution complicate this situation. It is noteworthy that official reports of dead otters in Chile are increasing, and there are no reports of dead otters in Peru. H5N1 avian influenza might be related to a significant increase in these reports by 2023.

Concrete conservation efforts include the establishment of protected natural areas such as the Paracas National Reserve, the only place where effective conservation has been pursued for several years, as shown in Chap. 5, "Conserving Marine Otter *Lontra felina* in One of the Oldest Marine Protected Areas of Southeast Pacific," by Raúl Sánchez-Scaglioni, Liliana Ayala, and Patricia Saravia-Guevara. However, they emphasize the need to organize and standardize procedures and protocols, following the details and recommendations provided by Medina-Vogel et al. in their chapters in this book. This will not only serve to improve conservation and biodiversity management in this protected area but also in other protected or unprotected areas from Peru to Chile.

The current conservation of the marine otter is the result of efforts initiated many years ago, detailed by Ricardo G. Correa, Johanna Garay-Rodríguez, and Raúl Sánchez-Scaglioni in Chap. 7, "Legal Framework and Its Application for the Conservation of the Marine Otter in Chile and Peru." Hunting of this species was gradually restricted until its total prohibition, and the international community joined agreements and conventions for the protection of wildlife. However, this does not prevent accidental, incidental, targeted, or pollution-related death. Therefore, it is essential to strengthen legal mechanisms and protocols for effective conservation.

This book also calls for research south of Chiloe Island, a great proportion of the marine otter distribution, but an area where there are no information of the species since the early 1980s.

We are very grateful to the chapter authors, as well as to Jaime Jahncke and Javier Matos for their chapter reviews. We appreciate the support of Springer Nature in publishing this book. We also thank editors Joao Pildervasser and Shanthini Kamaraj. Additionally, we wish to express our gratitude to our families, who supported and accompanied us on their adventures. Many thanks. To José, who is unfortunately no longer with us, we extend our heartfelt appreciation. Thank you, José.

Lima, Peru
Santiago, Chile

Liliana Ayala
Raúl Sánchez-Scaglioni
Gonzalo Medina-Vogel

This book takes a multidisciplinary approach to address conservation issues related to the marine otter (*Lontra felina*). The primary objective is to systematize various research efforts on this species, contributing to both the Peruvian and the Chilean otter action plan and the enhancement of conservation strategies for marine otters along its entire distribution. The authors share their accomplishments in conservation, ecology, and status in freshwater habitats, the effects of habitat fragmentation, interactions with human activities, and recommendations for the effective conservation of the species. This book is primarily directed at researchers, authorities, and individuals engaged in otter conservation efforts.

Contents

Contributors

Ulrich Zanabria-Alarcón Asociación para la Conservación y el Desarrollo Sostenible Sallqa Perú, Arequipa, Peru

Joanna Alfaro-Shigueto ProDelphinus, Lima, Peru

Centre for Ecology and Conservation, University of Exeter, Cornwall, UK

IUCN Species Survival Commission, Otter Specialist Group, Gland, Switzerland

Carrera de Biología Marina, Universidad Científica del Sur, Lima, Peru

Liliana Ayala Carrera de Biología Marina, Universidad Científica del Sur, Lima, Peru

Asociación Peruana para la Conservación de la Naturaleza, Lima, Peru

Carlos Calvo-Mac Centro de Investigación para la Sustentabilidad, Facultad Ciencias de la Vida, Universidad Andrés Bello, Santiago, Chile

IUCN Species Survival Commission, Otter Specialist Group, Gland, Switzerland

Proyecto Anzumo, ONG ConservAcción, Lima, Peru

Ricardo G. Correa IUCN Species Survival Commission, Otter Specialist Group, Valparaíso, Chile

Nicole Delgado-Parada Facultad de Ciencias Biológicas, Pontificia Universidad Católica de Chile, Santiago, Chile

Johanna Garay-Rodríguez Independent Researcher, Lima, Peru

Leonardo Hostos-Olivera Facultad de Ciencias y Filosofía/Centro de Investigación para el Desarrollo Integral y Sostenible, Laboratorios de Investigación y Desarrollo, Universidad Peruana Cayetano Heredia, Lima, Peru

Aldo G. Indacochea Grupo de Investigación Comunidades Acuáticas, Carrera de Biología Marina, Universidad Científica del Sur, Lima, Peru

Jeffrey C. Mangel ProDelphinus, Lima, Peru

Centre for Ecology and Conservation, University of Exeter, Cornwall, UK

IUCN Species Survival Commission, Otter Specialist Group, Gland, Switzerland

Gonzalo Medina-Vogel Centro de Investigación para la Sustentabilidad, Facultad Ciencias de la Vida, Universidad Andrés Bello, Santiago, Chile

IUCN Species Survival Commission, Otter Specialist Group, Gland, Switzerland

Gabriela Molina-Maldonado Facultad de Ciencias Biológicas, Pontificia Universidad Católica de Chile, Santiago, Chile

Clara Ortiz-Alvarez ProDelphinus, Lima, Peru

IUCN Species Survival Commission, Otter Specialist Group, Gland, Switzerland

Carrera de Biología Marina, Universidad Científica del Sur, Lima, Peru

Sea Mammal Research Unit, University of St Andrews, St Andrews, UK

Raúl Sánchez-Scaglioni Asociación Peruana para la Conservación de la Naturaleza, Lima, Peru

Patricia Saravia-Guevara Reserva Nacional de Paracas, Servicio Nacional de Áreas Naturales Protegidas por el Estado, Paracas, Peru

Joaquín A. Ugarte-Núñez Asociación para la Conservación y el Desarrollo Sostenible Sallqa Perú, Arequipa, Peru

Sonia Valle-Rubio Grupo de Investigación Comunidades Acuáticas, Carrera de Biología Marina, Universidad Científica del Sur, Lima, Peru

Juan Valqui Facultad de Ciencias y Filosofía/Centro de Investigación para el Desarrollo Integral y Sostenible, Laboratorios de Investigación y Desarrollo, Universidad Peruana Cayetano Heredia, Lima, Peru

Chapter 1
Environmental Components of the Marine Otter Habitat of Peru

Sonia Valle-Rubio and Aldo G. Indacochea

Abstract The habitat of the marine otter *Lontra felina* in Peru extends along the coastline from 9°S to 18°S within the Humboldt Current System. Oceanographic characteristics and coastal upwelling processes are responsible for cold temperatures in the subtropical coastal region of South America, creating a desert coastline with extensive sandy beaches interrupted by cliffs and rocky edges.

These environmental processes influence the biological diversity of the Peruvian Sea, with recurrent El Niño events affecting biodiversity. Biologically, intertidal communities on rocky shores provide a rich source of food for *L. felina*. The vast communities of macroalgae on rocky coastlines play an important role in food webs and as substrates for larval settlement of benthic organisms. In addition, they serve as refuge for benthic and pelagic species. Mollusks and crustaceans have high species richness values in coastal areas, and other invertebrates (polychaete annelids) are important groups in ecosystems and marine communities because of the diversity of their structures and functions. Marine vertebrate fauna encompasses a great diversity of fish, sea turtles, and up to 30 mammals. Hundred species of seabirds, many of which are coastal, are noticeable for their high population densities.

The Peruvian Sea faces threats such as climate change, pollution, and illegal fishing practices. In recent years, progress has been made in marine conservation and its resources in Peru. Currently, a holistic perspective involves conserving species and their marine environments with the creation of new protected marine natural areas.

Keywords Habitat · Rocky shores · Peruvian coastline · Rocky zonation · Humboldt current system

---

S. Valle-Rubio (✉) · A. G. Indacochea
Grupo de Investigación Comunidades Acuáticas, Carrera de Biología Marina, Universidad Científica del Sur, Lima, Peru
e-mail: svalle@cientifica.edu.pe; aindacocheam@cientifica.edu.pe

L. Ayala et al. (eds.), *Marine Otter Conservation*,
https://doi.org/10.1007/978-3-031-53931-2_1

1.1 Introduction

The Peruvian coastline extends from north to south, from the Ecuadorian border at Punta Capones (03°23′ 31.6″ S, 80°18′ 49.2″ W) to the Chilean border at Punto Concordia (18°21′ 00.4″S, 70°22′ 49.8″ W), covering a total length of 3079.5 km (IGN, 2020). The habitat of the marine otter *Lontra felina* in Peru spans from 8°S as the northern limit to 18°S (Santillán & Caro, 2007; Alfaro-Shigueto et al., 2011; Apaza & Romero, 2012) in ecosystems within the extensive Humboldt Current System (HCS), that extends along the coasts of Peru and Chile, between 4°S and 45°S (Chavez & Messié, 2009).

The HCS is a coastal marine current that flows from south to north at speeds between 4 and 15 cm/s from the surface to a maximum depth of 200 m (Tarazona & Arntz, 2001). Because of its geographical position on the eastern edge of the Pacific Ocean, the HCS covers a large area of coastal upwelling (Sydeman et al., 2014). Coastal upwelling is the upward movement of subsurface seawater that carries nutrients and cooler water to the surface layer (Barber, 2001). Coastal upwelling off the Peruvian coast occurs throughout the year with some temporal variability, which affects the abundance and distribution of certain hydrobiological resources (Graco et al., 2007).

Coastal upwelling processes result from Ekman transport (Ekman, 1905). This theory has contributed to the understanding of upwelling processes, which are responsible for the low temperature in the subtropical coastal region of South America and the desert coastline (Petersen & Mugica Martinez, 1972; Arntz & Eberhard, 1996). The desert coastline, with its extensive sandy beaches, is a dynamic ecosystem that provides habitats for diverse macrobenthic fauna (McLachlan & Dorvlo, 2005). It is interrupted by areas of cliffs and rocky edges (Sandweiss et al., 1996), which are the preferred habitats of *L. felina* (Pizarro, 2021).

L. felina inhabits the marine-coastal zone of Peru and Chile, particularly in coastal ecosystems in shore zones and rocky cliffs (Alfaro-Shigueto et al., 2011). Marine otters can also utilize sandy beach areas, as suggested by footprints sighted in central coastal Peru (Paracas, Ica), consistent with reports from other latitudes (Rozzi & Torres-Mura, 1990).

In this review, coastal ecosystems will be considered as those in which biological processes occur, involving the interaction between biotic and abiotic factors, and the flow of energy within the coastal marine zones of the habitat of *L. felina*. Subsequently, we provide an overview of the characteristics and variability of biotic and abiotic factors in the coastal marine zones of Peru from an ecological perspective. This allows the features of homogeneity and predictable patterns of variability to be captured over time.

1.2 The Peruvian Coast

1.2.1 *Description of the Geology and Coastal Topography of Peru*

Along the coastline exposed to the Pacific Ocean, the intertidal zone is highly exposed to air and seawater, whereas the subtidal zone is always submerged (Rigby et al., 2007). These areas encompass the geomorphological space on either side of the seashore, where marine and continental parts interact. This transition between rocky coastal shores and the terrestrial environment is not abrupt but gradual, primarily because of water movement associated with waves and tides (Chappuis et al., 2014).

The subtidal zone is a continental shelf (CS), limited to the east by the Peruvian coast and to the west by the oceanic Nazca plate, extending from where a steep slope change occurs between 100 and 200 m in depth (Morales et al., 2020). Relatively rapid marine sedimentation occurs in this area, which is characteristic of shallow waters (Petersen & Mugica Martinez, 1972). The width of the CS varies depending on the geodynamics of the continental margin (Schweigger, 1964). It is divided into the northern zone (Tumbes to Punta Aguja, Piura), where the CS is relatively narrow (6–65 km); the central zone (Punta Aguja and Isla San Gallán), where the CS is wider, between 9 and 130 km; and the southern zone, where the 160 m isobath is close to the coastline, and the CS width is only between 2 and 4 km (Bruland et al., 2005; Morales et al., 2020) (Fig. 1.1).

In general, the western edge of South America corresponds to paleo-rocky coasts interrupted by paleo-sedimentary systems (Blanco-Chao et al., 2014). Currently, the western coast of South America is dominated by rocks (Inman & Nordstrom, 1971).

One of the main factors for *L. felina* to choose its burrow establishment site seems to be the availability of suitable galleries among rocks, and these types of spaces are found in coastal regions, mainly in rocky cliffs (Sielfeld & Castilla, 1999). *L. felina* is distributed along Peruvian coast in three geographical areas, differentiated by their geomorphology, and where the occurrence of steep places and suitable natural rocky gallery formations for *L. felina* burrows would be more probable (Apaza & Romero, 2012). The distribution areas of *L. felina* in Peru would then be from south to north: from 18°S to 13°47′S; a second unit from 13°47′S to 10°S; and the northernmost distribution unit between 08°58′S and 05°47′S (Apaza & Romero, 2012). These three geomorphological units are within the biogeographic region or province of Humboldt (Ibanez-Erquiaga et al., 2018).

The coastline unit from 18°S to 13°47′S is mostly comprised of marine terraces (Apaza & Romero, 2012). They are relatively flat surfaces of marine origin, bounded by a steeper ascending slope on one side and a steeper descending slope on the opposite side (see https://portal.ingemmet.gob.pe/web/guest/mapa-geomorfologico). Marine terraces can result from marine abrasion or denudation, or they may consist of accumulations of materials removed by coastal erosion from shallow to slightly emergent waters (constructed marine terraces) or may also have a polygenic origin,

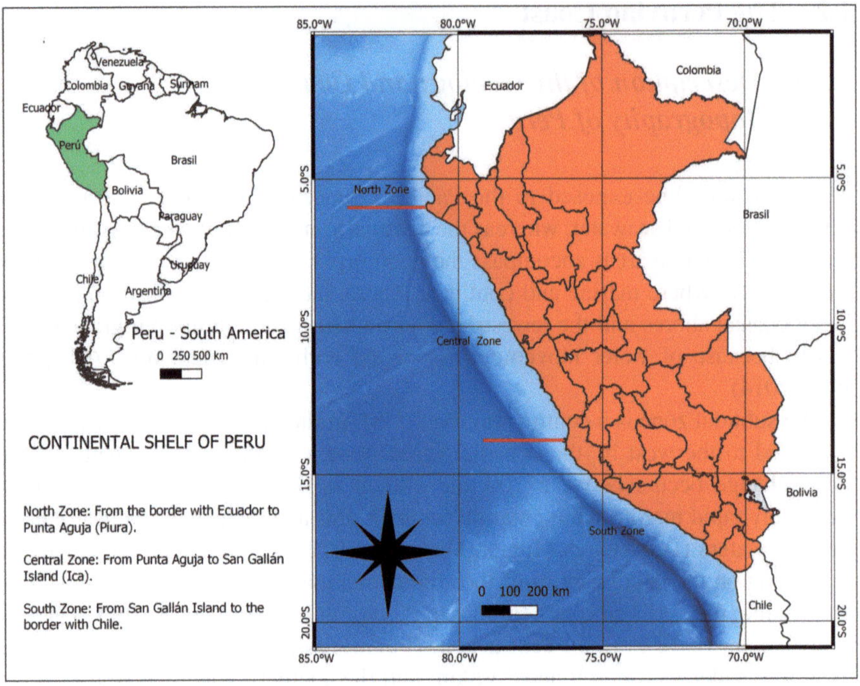

Fig. 1.1 Extension of the continental shelf of Peru

with the occurrence of in situ deposits after an abrasion phase. That area also has extensive cliffs of metamorphic and intrusive rocks, such as those found at the ends of Independence Bay in the department of Ica (Fig. 1.2). Biologically, the intertidal communities of rocky shores in the southern region of Humboldt Province are dominated by macroalgae, and despite having lower richness than the northern zone, they contain the highest abundances (Ibanez-Erquiaga et al., 2018).

An intermediate strip constitutes the second geomorphological unit of *L. felina* sightings, ranging from 13°47′S to approximately 10°S (Apaza & Romero, 2012). Geologically, this intermediate coastal strip is composed of cliffs, hills, and mounds of intrusive rocks that are interspersed with marine terraces. Further north in the Ica Department, there is still a long coastal strip in an alluvial plain (13°47′S–13°23′S). There are extensive marine terraces up to the Lima Department with dune fields along its northern course. Between northern Lima and Chimbote (11°50′S–09°S), the geomorphology is highly varied, with strips of sand blankets, alluvial plains, coastal wetlands, and interspersed strips of volcanic or sedimentary rock hills and sand blankets (Apaza & Romero, 2012).

Despite the presence of large areas that may not be suitable for otter settlement in this "central strip," the biological diversity of rocky areas provides a good food supply for *L. felina*. Paredes and Tarazona (1980) described the diversity of organisms on the coast of the Lima Department associated with clusters of mussels

Fig. 1.2 The coastal area of the Paracas National Reserve (Ica) is formed by terraces of metamorphic and intrusive rocks. (Photo: Aldo Indacochea)

inhabiting the mid-littoral rocky zone, formed by the *Perumytilus purpuratus* and *Semimytilus patagonicus* communities.

The third geomorphological unit of *L. felina* sightings is found in the northern zone of the Peruvian province from 05°47′S (Apaza & Romero, 2012). The coastline consists of a rocky cliff promontory (see: https://portal.ingemmet.gob.pe/web/guest/mapa-geomorfologico), it is an area of very high species richness (Ibanez-Erquiaga et al., 2018). However, on southward, a long stretch of sand blankets and floodplains dominates until 08°58′S, which is the southern limit of the third otter sighting unit.

1.2.2 *The Biological Diversity of the Peruvian Sea Coast*

The geographical configuration of the Peruvian coastline is responsible for its high productivity. The areas of bays and coves harbored the greatest diversity of species and densities, mainly invertebrates. These areas are also important because they support recruitment and serve as refuges for many species; some of them undertake seasonal migrations to these areas for spawning (e.g., anchovy and common squid) (Argüelles et al., 2010).

The biological diversity of the Peruvian Sea is conditioned by various factors, such as its geological origin, latitudinal temperature gradient, direction of marine currents, coastal upwellings, and a particular aspect of biodiversity—the effects of recurrent El Niño (Tarazona & Valle, 1998).

The coastline, in general, and rocky shores, in particular, are areas with high biodiversity (Tarazona & Valle, 1998). They present the most adverse physical conditions of all marine environments, resulting in a strong gradient of environmental stress perpendicular to the coast related to desiccation, temperature, and irradiance, being more extreme in the supralittoral zone (Chappuis et al., 2014).

The upper distribution limit of species in the rocky intertidal zone is determined by their tolerance to abiotic stress (Stickle et al., 2017). This gradient of environmental "stress" generates a pattern of organism distribution called littoral zonation in which species replace each other in bands or zones from the lower limit of the lowest tide to the upper limit of the highest tide (Stephenson & Stephenson, 1949; Paredes, 1974). The upper limits of species are defined by their physiological tolerance to emersion, whereas the lower limits are defined by their ability to compete for space, obtain food, and avoid potential predators (Southward, 1958).

Studies on the zonation patterns of rocky shores in the Peruvian Littoral zone are relatively scarce. In 1970s, the first models of rocky shore zonation were obtained for the central and southern Peruvian coast (Paredes, 1974; Paredes & Tarazona, 1980; Tejada-Pérez et al., 2018). Paredes (1974) establishes the zoning scheme on the solid rocky shore of the Lima Department.

The rocky zonation of Paredes (1974) best represents the biological diversity of the ecosystems in Peru. It is limited to the upper edge of the supralittoral strip in the presence of the gastropod *Echinolittorina peruviana*, although it is not considered the key species, as its presence and distribution depend on the location of the zone, whether it is a protected area or is more exposed to wave action (Fig. 1.3). In this rocky zonation strip, there is also the rock crab *Grapsus*, which is present on all shores analyzed in this study, and the gastropod *Scurria viridula*, both of which are reported as part of the diet of *L. felina* (Biffi & Iannacone, 2010).

The mid-littoral zone is inhabited by barnacle *Jehlius cirratus* and kelp algae, with a community featuring the bivalve *Perumytilus purpuratus* as the most representative key species. This mid-littoral characterization of *P. purpuratus* is found in areas further south of Peru (Tejada-Pérez et al., 2018) and dominates the rocky intertidal substrates in Chile (Castilla et al., 2014). The *P. purpuratus* community is located from the mid-littoral in protected shores to the upper mid-littoral in highly exposed shores, presenting a fauna associated with several species considered favored prey of *L. felina* (barnacles, chitons, and the bivalves) (Biffi & Iannacone, 2010). Meanwhile, the *S. patagonicus* community occupies the lower mid-littoral horizon, with invertebrate species such as polychaetes, gastropod mollusks such as *Tegula atra*, and some algae species (Paredes, 1974).

The lowest part is the infralittoral zone, which is always submerged, where another larger barnacle species, *Austromegabalanus psittacus*, is found, along with forests of *Lessonia* spp. and *Macrocystis pyrifera* (Paredes, 1974).

Overall, there is greater species richness at lower latitudes, higher abundance at southern latitudes, and marked variability between these biogeographic units (Ibanez-Erquiaga et al., 2018).

Fig. 1.3 Zonation pattern on the rocky shore of the central coast of Peru. Guano birds use the rocky supralittoral zone as a resting area. (Photo: Sonia Valle)

The intertidal and subtidal coastal zones of Peru harbor large communities of primary producers, such as macroalgae, highlighting their importance in the ecosystem because of their role in trophic chains (Arakaki et al., 2018). Carbon storage is one of the most important ecosystem services provided by macroalgal beds (Aller-Rojas et al., 2020). Macroalgae serve not only as a substrate for the larval settlement of benthic organisms, but also as a refuge for benthic and pelagic species, including some fish as part of the diet of *L. felina* (Fig. 1.4) (Biffi & Iannacone, 2010, see Chapter 3).

An updated list of marine macroalgae in Peru reported a total of 260 species, such as Rhodophyta (185 species), Chlorophyta (41 species), and Phaeophyceae (34 species) (Avila-Peltroche & Padilla-Vallejos, 2020).

The biocenoses of the Peruvian littoral zone, particularly those of rocky shores, are known for their high diversity, which is attributed to the rocky substrates that provide the necessary hard substrate for the settlement of sessile organisms, especially macroalgae. This habitat also supports mussel communities such as the *Perumytilus purpuratus* community, which can consist of up to 87 species, and the *Semimytilus patagonicus* community, which can harbor up to 77 species (Tarazona & Valle, 1998).

Intertidal mussels are considered "ecosystem engineers" because they host a high biodiversity of invertebrates (Arribas et al., 2014). The empty shells and threads formed through the byssus gland as well as the interstitial space create a substrate and habitat for colonization and refuge for the reproductive and juvenile

Fig. 1.4 A brown macroalgae forest (*Lessonia trabeculata*) and a "Pintadilla" (*Cheilodactylus variegatus*) swim in the foreground. (Photo: Bernabé Moreno)

Fig. 1.5 Gastropod mollusk *Thaisella chocolata* and egg masses on a bed of mussels provide habitat for other organisms. (Photo: Aldo Indacochea)

stages of other invertebrate species, mostly macrofauna (Fig. 1.5) (Thiel & Ullrich, 2002). Additionally, mussel "beds" also trap sediment and organic particles that serve as food for small invertebrates (Arribas et al., 2014).

Mollusks are the most diverse group of organisms in coastal zones, occupying various habitats with different substrate types, such as rocks, boulders, angular pebbles, sand, and mud (Montes-Iturrizaga, 2018). In Peru, there are records of 1088 mollusk species, mainly gastropods (570 species) and bivalves (370 species) (Paredes, 2012; cited in Montes-Iturrizaga, 2018). The highest number of mollusk species in the Peruvian Sea is distributed in the tropical waters of the Panamanian Province north of 05°40′S, with lower diversity in the Peruvian Province area (Ramírez et al., 2003).

The dominance of mollusk species in intertidal zones is also associated with the biogeographic regions. In the southern coastal zone (17°S), a small-sized gastropod species, *Echinolittorina peruviana*, dominates mainly in the supralittoral fringe, while *Perumytilus purpuratus* dominates the mid-littoral zone, and the gastropod *Tegula atra* dominates in the infralittoral fringe (Tejada-Pérez et al., 2018; Valqui et al., 2021). In the central zone of the Humboldt Province, *P. purpuratus* dominates, while in the northern zone, the dominant species in rocky intertidal areas is the gastropod *Stramonita haemastoma* (Valqui et al., 2021), a predatory snail.

Choromytilus chorus "chorozapato" is the largest species in the mussel family and is more common in southern latitudes (although its distribution extends from 7° South). Its habitat is rocky mid-littoral and infralittoral between 4 and 20 m depth on hard substrates (Montes-Iturrizaga, 2018) and represents a potential resource in the otter's diet.

Crustaceans also constitute an important group of invertebrates in the Peruvian sea. A total of 361 species and 216 genera of marine decapod crustaceans and 15 stomatopods have been recorded (Moscoso, 2012). Additionally, 24 species of cirripeds have been documented (Tarazona et al., 2003), of which barnacles are the major prey for *L. felina* (Biffi & Iannacone, 2010).

Alongside mollusks and crustaceans, polychaete annelids constitute crucial groups in marine ecosystems owing to their diverse structures and functions (Tarazona, 1974), and they represent the most abundant taxon in benthic communities. Up to 188 species of polychaetes from 113 genera and 43 families have been recorded (Aguirre & Canales, 2017).

Echinoderms are representative groups of rocky, sandy, and subtidal communities. Echinoderms play significant roles in ecosystems, such as oxygenation of sandy or muddy bottoms, control of macroalgae by feeding on them, and maintaining their growth in check, and they serve as food for many animals, including *L. felina* (Sanino & Meza, 2016). Hooker et al. (2013) report 215 species of echinoderms: 1 Crinoidea species, 64 Asteroidea species, 42 Ophiuroidea species, 35 Echinoidea species, and 73 Holothuroidea species (Fig. 1.6).

Sponges are other important invertebrates in littoral and subtidal ecosystems, with 43 reported species. Sponges play a role in ecosystem formation by providing refuge for many other invertebrates (Pearse, 1934; Frith, 1976; Kunzmann, 1996).

Marine vertebrate fauna, including the neritic zone, encompasses a wide diversity of fish, with 1070 reported species (Tarazona et al., 2003). Five species of sea turtles (*Dermochelys coriacea*, *Chelonia mydas*, *Lepidochelys olivacea*, *Eretmochelys imbricata*, and *Caretta caretta*) are found in the Peruvian sea (Hays-Brown et al., 1982). Either during their migratory journeys, for feeding, or as a habitat for the development of juveniles (Rosales et al., 2010).

Fig. 1.6 Sea urchins of the species *Arbacia spatuligera* in the subtidal zone of Pucusana, Lima, on the central coast of Peru. (Photo: Aldo Indacochea)

Fig. 1.7 Edge of Asia Island (Lima, Peru) with a resting area for sea lions (*Arctocephalus australis*) sharing space with guano birds (*Pelicanus thagus*). (Photo: Sonia Valle)

Up to 30 species of marine mammals have been reported, in addition to *Lontra felina*: 3 species of sea lions, 5 species of baleen whales, 6 species of toothed whales, 1 species of Phocoenidae, and 13 species of dolphins (Majluf & Reyes, 1989) (Fig. 1.7).

Fig. 1.8 The Humboldt Penguin *Spheniscus humboldti* (Photo: Aldo Indacochea) and the Inca Tern *Larosterna inca.* (Photo: Sonia Valle)

Up to 100 species of seabirds are known, some of which stand out for their high population densities and guano formation, including *Spheniscus humboldti*, *Pelecanoides garnotii*, *Sula variegata*, *Sula nebouxii*, *Phalacrocorax bougainvillii*, *Phalacrocorax gaimardi*, *Pelecanus thagus*, and *Larosterna inca* (Goya Sueyoshi et al., 2016) (Fig. 1.8).

The geographical configuration of the Peruvian coast is responsible for its high productivity. The bay and inlet areas had the highest species diversity and density, primarily invertebrates. These areas are also important recruitment sites that serve as refuges for many species. Some species undertake seasonal migration to these areas for spawning, such as the pejerrey *Odontesthes regia* and the common squid *Doryteuthis gahi* (Argüelles et al., 2010).

1.2.3 Threats to Coastal Marine Biodiversity

The Peruvian Sea faces challenges like those affecting oceans worldwide. These include threats of climate change, such as rising sea temperatures, acidification, sea-level rise, decreased oxygen availability, pollution, and illegal fishing practices. These threats have led to a loss of marine biodiversity (Pan et al., 2013).

Currently, we have more knowledge about the consequences of climate change and global warming and the effect of sea level rise in coastal areas is an environmental factor to consider. However, zonation (the distribution of organisms in parallel bands or zones) does not result from tidal action, as it occurs both below and above the water level around "static" waters where there is no change due to tides (Stephenson & Stephenson, 1949). According to the IPCC (2019), the global sea-level rise between 1902 and 2015 averaged 0.17 m. In Peru, sea-level changes of up to 200 mm/year have been recorded during "El Niño" events (Jigena-Antelo et al., 2023). Nevertheless, Tavares Corrêa (2022) reported an increase in the mean sea level did not have a major influence on beach erosion in Peru.

The Peruvian Sea is highly productive, with extensive areas exhibiting low oxygen levels owing to high demand combined with poor ventilation and slightly low pH (Graco et al., 2007). These oceanic areas are known as Oxygen Minimum Zones (OMZ), and the OMZ in the Peruvian Sea is the shallowest of global oceans (Helly & Levin, 2004). Coastal areas ranging from the high tide limit to a depth of approximately 30–40 m are affected by hypoxia events and oceanographic changes associated with El Niño (Tarazona et al., 2003). In addition to hypoxia, other environmental factors in the coastal zone can also influence the physiology, metabolism, and structure of organism communities (Cruz-Motta et al., 2020). Thereby affecting the seasonal availability of prey organisms for *L. felina* (Biffi & Iannacone, 2010).

Regarding pollution in the Peruvian Sea, the major identified sources include domestic and industrial waste, petroleum hydrocarbons, and chlorinated hydrocarbons (Guillén et al., 1980). The most significant accumulation of marine litter is undoubtedly constituted by plastic materials from domestic and industrial wastes, which are directly and indirectly discharged into the ocean. The consequence of this is a reduction in habitat quality (benthic and pelagic) and a high risk of ingestion by organisms.

The damage caused by large plastics, macroplastics, as well as microplastics (particles smaller than 5 mm), to marine life extends from plankton and benthic invertebrates to large mammals. Microplastics and the toxins they carry can be consumed by low trophic level organisms and pass up the food chain, for example, from plankton to fish and mammals, in bioaccumulation processes (Parolini et al., 2023). Santillán et al. (2020) reported microplastics in the feces of L. *felina*, probably resulting from bioaccumulation through the consumption of prey such as pejerrey (*Odontesthes regia*) and pintadilla (*Cheilodactylus variegatus*).

In recent years, progress has been made in marine and resource conservation in Peru. Currently, there is a more holistic approach involving the conservation or protection of species and the marine environment in which they thrive (UNEP, 2011). In Peru, there are three marine protected areas: the San Fernando National Reserve, Paracas National Reserve, and the National Reserve of the Islands, Islets, and Guano Points System, the most recent of which was created in 2011 (SERNANP, 2016) (Fig. 1.9). These areas contribute to the conservation of marine biodiversity; however, it is essential to regulate the use and transformation of the coastline, as well as the extraction of resources and artisanal fishing in unprotected areas.

Acknowledgement The authors would like to thank Bernabé Moreno for providing the Fig. 1.4 during the execution of the project "Determination of the distribution, biomass, and ecological importance of the 'aracanto palo' *Lessonia trabeculata* in the marine environment of the San Fernando National Reserve," funded by KfW Development Bank, Peruvian Trust Fund for National Parks and Protected Areas (PROFONANPE), the National Service of Natural Protected Areas (SERNANP), and the Consortium for Economic and Social Research (CIES).

Fig. 1.9 Marine protected areas in Peru: Paracas National Reserve, Ica Department, Peru (**a**); Asia Island, Lima, representing the National Reserve of the Islands, Islets, and Guano Points System (**b** and **c**); and San Fernando National Reserve, Arequipa Department (**d**)

References

Aguirre, L., & Canales, R. (2017). Poliquetos de Perú: Estado actual y perspectivas para la investigación. 101-114. En: Díaz-Díaz, O., D. Bone, C. T. Rodríguez & V. H. Delgado-Blas (Eds.) 2017. Poliquetos de Suidamerica. Volumen especial del Boletín del Instituto Oceanográfico de Venezuela. Cumaná, Venezuela, 149 p.

Alfaro-Shigueto, J., Valqui, J., & Mangel, J. C. (2011). Nuevo registro de la nutria marina Lontra felina (Molina, 1782) al norte de su distribución actual. *Ecología Aplicada, 10*, 87–91.

Aller-Rojas, O., Moreno, B., Aponte, H., & Zavala, J. (2020). Carbon storage estimation of Lessonia trabeculata kelp beds in southern Peru: An analysis from the San Juan de Marcona region. *Carbon Management, 11*, 525–532. https://doi.org/10.1080/17583004.2020.1808765

Apaza, M., & Romero, L. (2012). Distribución y observaciones sobre la población de la nutria marina Lontra felina (Molina 1782) en el Perú. *Revista Peruana de Biología, 19*, 285–298. https://doi.org/10.15381/rpb.v19i3.1064

Arakaki, N., Gil-Kodaka, P., Carbajal, P., et al. (2018). I-rhodophyta. In *Macroalgas de la Costa Central del Perú* (p. 126). UNALM.

Argüelles, J., Taipe, A., & Espinoza, E., et al. (2010). *Delimitación y caracterización de bancos naturales de invertebrados bentónicos comerciales y áreas de pesca artesanal en la región Lima entre Punta Litera – Playa Grande* (Barranca y Huara de la Región Lima).

Arntz, W., & Eberhard, F. (1996). *El niño: experimento climático de la naturaleza causas físicas y efectos biológicos*. Fondo de Cultura Económico.

Arribas, L. P., Donnarumma, L., Palomo, M. G., & Scrosati, R. A. (2014). Intertidal mussels as ecosystem engineers: Their associated invertebrate biodiversity under contrasting wave exposures. *Marine Biodiversity, 44*, 203–211. https://doi.org/10.1007/s12526-014-0201-z

Avila-Peltroche, J., & Padilla-Vallejos, J. (2020). The seaweed resources of Peru. *Botanica Marina, 63*, 381–394. https://doi.org/10.1515/bot-2020-0026
Barber, R. T. (2001). Upwelling ecosystems. In *Encyclopedia of ocean sciences* (pp. 3128–3135). Elsevier.
Biffi, D., & Iannacone, J. (2010). Variabilidad trófica de Lontra feline (Molina 1782) (Carnivora: Mustelidae) en dos poblaciones de Tacna (Perú) entre Agosto y Diciembre de 2006. *Mastozoología Neotropical, 17*, 11–17.
Blanco-Chao, R., Pedoja, K., Witt, C., et al. (2014). The rock coast of South and Central America. In *Geological society memoir* (pp. 155–191). Geological Society of London.
Bruland, K. W., Rue, E. L., Smith, G. J., & DiTullio, G. R. (2005). Iron, macronutrients and diatom blooms in the Peru upwelling regime: Brown and blue waters of Peru. *Marine Chemistry, 93*, 81–103. https://doi.org/10.1016/j.marchem.2004.06.011
Castilla, J. C., Yanez, E., Silva, C., & Fernandez, M. (2014). A review and analysis of Easter Islands traditional and artisan fisheries. *Latin American Journal of Aquatic Research, 42*, 690–702. https://doi.org/10.3856/vol42-issue4-fulltext-3
Chappuis, E., Terradas, M., Cefalì, M. E., et al. (2014). Vertical zonation is the main distribution pattern of littoral assemblages on rocky shores at a regional scale. *Estuarine, Coastal and Shelf Science, 147*, 113–122. https://doi.org/10.1016/j.ecss.2014.05.031
Chavez, F. P., & Messié, M. (2009). A comparison of eastern boundary upwelling ecosystems. *Progress in Oceanography, 83*, 80–96. https://doi.org/10.1016/J.POCEAN.2009.07.032
Cruz-Motta, J. J., Miloslavich, P., Guerra-Castro, E., et al. (2020). Latitudinal patterns of species diversity on South American rocky shores: Local processes lead to contrasting trends in regional and local species diversity. *Journal of Biogeography, 47*, 1966–1979. https://doi.org/10.1111/jbi.13869
Ekman, V. W. (1905). On the influence of the earth's rotation on ocean currents. *Archive for Mathematics, Astronomy and Physics, 2*, 1–52.
Frith, D. W. (1976). Animals associated with sponges at North Hayling, Hampshire. *Zoological Journal of the Linnean Society, 58*, 353–362. https://doi.org/10.1111/j.1096-3642.1976.tb01005.x
Goya Sueyoshi, E., Romero Moreno, C., Villagra Petre, D., et al. (2016). *Aves Marinas en el Perú*. IMARPE.
Graco, M. I., Ledesma, J., Flores, G., & Girón, M. (2007). Nutrientes, oxígeno y procesos biogeoquímicos en el sistema de surgencias de la corriente de Humboldt frente a Perú. *Revista Peruana de Biología, 14*, 117–128. https://doi.org/10.15381/rpb.v14i1.2165
Guillén, O., Asthu, V., & Aquino, R. (1980). *Contaminación marina en el Perú*. Instituto del Mar del Perú.
Hays-Brown, C., Brown, W. M., & Bjorndal, K. A. (1982). Status of sea turtles in the southeastern Pacific: Emphasis on Peru. In *Biology and conservation of sea turtles* (pp. 235–240). Smithsonian Press.
Helly, J. J., & Levin, L. A. (2004). Global distribution of naturally occurring marine hypoxia on continental margins. *Deep Sea Research Part I: Oceanographic Research Papers, 51*, 1159–1168. https://doi.org/10.1016/j.dsr.2004.03.009
Hooker, Y., Prieto-Rios, E., & Solís-Marín, F. A. (2013). Echinoderms of Peru. In *Echinoderm research and diversity in Latin America* (pp. 277–299). Springer-Verlag.
Ibanez-Erquiaga, B., Pacheco, A. S., Rivadeneira, M. M., & Tejada, C. L. (2018). Biogeographical zonation of rocky intertidal communities along the coast of Peru (3.5–13.5 S Southeast Pacific). *PLoS One, 13*, e0208244. https://doi.org/10.1371/journal.pone.0208244
IGN. (2020). *Atlas Digital del Perú*. https://app4.ign.gob.pe/bienvenidos.php. Accessed 25 Dec 2023.
Inman, D. L., & Nordstrom, C. E. (1971). On the tectonic and morphologic classification of coasts. *Journal of Geology, 79*, 1–21. https://doi.org/10.1086/627583
IPCC, (2019). IPCC Special Report on the Ocean and Cryosphere in a Changing Climate [H.-O. Pörtner, D.C. Roberts, V. Masson-Delmotte, P. Zhai, M. Tignor, E. Poloczanska,

K. Mintenbeck, A. Alegría, M. Nicolai, A. Okem, J. Petzold, B. Rama, N.M. Weyer (eds.)]. https://www.ipcc.ch/site/assets/uploads/sites/3/2019/12/SROCC_FullReport_FINAL.pdf

Jigena-Antelo, B., Estrada-Ludeña, C., Howden, S., et al. (2023). Evidence of sea level rise at the Peruvian coast (1942–2019). *Science of the Total Environment, 859*, 160082. https://doi.org/10.1016/j.scitotenv.2022.160082

Kunzmann, K. (1996). Associated fauna of selected sponges (Hexactinellida and Demospongiae) from the Weddell Sea, Antarctica. *Berichte zür Polarforschung, 210*, 1–93.

Majluf, P., & Reyes, J. C. (1989). The marine mammals of Peru: A review. In: *The Peruvian upwelling ecosystem. Proc. workshop on models for yield prediction, Callao, 1987*, pp. 344–363.

McLachlan, A., & Dorvlo, A. (2005). Global patterns in sandy beach macrobenthic communities. *Journal of Coastal Research, 21*, 674–687. https://doi.org/10.2112/03-0114.1

Montes-Iturrizaga, D. (2018). *Guía para la identificación de Gasterópodos y bivalvos marinos del Perú*. Fondo Editorial de la Universidad Científica del Sur. 382 p.

Morales, M., Cerpa, L., Cornejo, T., et al. (2020). *Geología de la plataforma continental del Perú: paralelos 03°30′ y 14°00′ latitud sur*. Instituto Geológico, Minero y Metalúrgico – INGEMMET.

Moscoso, V. (2012). *Catálogo de crustáceos decápodos y estomatópodos del Perú*. Boletín del Instituto del Mar del Perú, 27(1–2), 8–207. Lima.

Pan, J., Marcoval, M. A., Bazzini, S. M., et al. (2013). Coastal marine biodiversity challenges and threats. In *Marine ecology in a changing world* (pp. 43–67). CRC Press.

Paredes, C. (1974). El modelo de zonación en la orilla rocosa del departamento de Lima. *Revista Peruana de Biología, 1*, 168–191. https://doi.org/10.15381/rpb.v1i2.8422

Paredes, C. Q., & Tarazona, J. B. (1980). Las comunidades de mitilidos del mediolitoral rocoso del departamento de Lima. *Revista Peruana de Biología, 2*, 59–71. https://doi.org/10.15381/rpb.v2i1.8362

Paredes C. (2012). Estado del arte de la taxonomía de moluscos marinos en el Perú (videoconferencia). En: Semana Nacional de la Diversidad Biológica Marina Costera. Una mirada al mar que no conocemos. Lima: Ministerio del Ambiente 2012.

Parolini, M., Stucchi, M., Ambrosini, R., & Romano, A. (2023). A global perspective on microplastic bioaccumulation in marine organisms. *Ecological Indicators, 149*, 110179. https://doi.org/10.1016/j.ecolind.2023.110179

Pearse, A. S. (1934). Inhabitants of certain sponges at Dry Tortugas. *Carnegie Institute of Washington, Papers of Tortugas Laboratory, 28*, 117–124..

Petersen, G., & Mugica Martinez, R. (1972). *Historia Marítima del Perú. El mar gran personaje. Geografía y geología general del litoral peruano*. Instituto de Estudios Históricos – Marítimos del Perú.

Pizarro, J. (2021). *La Nutria marina en el sur del Perú*. Asociación para el Desarrollo de las Ciencias Biológicas en el Perú.

Ramírez, R., Paredes, C., & Arenas, J. (2003). Moluscos del Perú. *Revista de Biología Tropical, 51*, 225–284.

Rigby, R., Iken, K., & Kato, T. (2007). *Sampling biodiversity in coastal communities NaGISA protocols for seagrasses and macroalgal habitats*. NUS Press.

Rosales, C. A., Vera, M., & Llanos, J. (2010). Varamientos y captura incidental de tortugas marinas en el litoral de Tumbes, Perú. *Revista Peruana de Biología, 17*. https://doi.org/10.15381/rpb.v17i3.4

Rozzi, R., & Torres-Mura, J. C. (1990). Observaciones del chungungo (Lutra felina) al sur de la isla Grande de Chiloé: antecedentes para su conservación. *Medio Ambient, 11*, 24–28.

Sandweiss, D. H., Richardson, J. B., Reitz, E. J., et al. (1996). Geoarchaeological evidence from Peru for a 5000 years B.P. onset of El Nino. *Science (80-), 273*, 1531–1533. https://doi.org/10.1126/science.273.5281.1531

Sanino, G. P., & Meza, M. I. (2016). Ecología Trófica y Simpatría de Nutrias (Lontra felina y Lontra provocax) en la Reserva Añihué, Patagonia Chilena. *Boletín Museo Nacional de Historia Natural, 65*, 279–289. https://doi.org/10.54830/bmnhn.v65.2016.98

Santillán, L., & Caro, K. (2007). Mamíferos Marinos en el Puerto de Salaverry y Zonas Adyacentes. In: *1er Congreso Nacional de Ciencias del Mar del Perú*.

Santillán, L., Saldaña-Serrano, M., & De-La-Torre, G. E. (2020). First record of microplastics in the endangered marine otter (Lontra felina). *Mastozoología Neotropical, 27*, 211–215. https://doi.org/10.31687/saremMN.20.27.1.0.12

Schweigger, E. (1964). *El litoral Peruano* (2nd ed.). Gráfica Morsom S.A.

SERNANP. (2016). *Áreas Naturales Protegidas del Perú (2011–2015). Conservación para el desarrollo sostenible*. MINAM.

Sielfeld, W., & Castilla, J. C. (1999). Estado de conservación y conocimiento de las nutrias en Chile. *Estudios Oceanologicos, 18*, 69–79.

Southward, A. J. (1958). Note on the temperature tolerances of some intertidal animals in relation to environmental temperatures and geographical distribution. *Journal of the Marine Biological Association of the United Kingdom, 37*, 49–66. https://doi.org/10.1017/S0025315400014818

Stephenson, T. A., & Stephenson, A. (1949). The universal features of zonation between tide-marks on rocky coasts. *Journal of Ecology, 37*, 289. https://doi.org/10.2307/2256610

Stickle, W. B., Carrington, E., & Hayford, H. (2017). Seasonal changes in the thermal regime and gastropod tolerance to temperature and desiccation stress in the rocky intertidal zone. *Journal of Experimental Marine Biology and Ecology, 488*, 83–91. https://doi.org/10.1016/j.jembe.2016.12.006

Sydeman, W. J., García-Reyes, M., Schoeman, D. S., et al. (2014). Climate change and wind intensification in coastal upwelling ecosystems. *Science (80-), 345*, 77–80. https://doi.org/10.1126/science.1251635

Tarazona, J. (1974). Lista de poliquetos sedentarios hallados en el Peru. *Revista Peruana de Biología, 1*, 164–167. https://doi.org/10.15381/rpb.v1i2.8419

Tarazona, J., & Arntz, W. (2001). The Peruvian coastal upwelling system. In U. Seeliger & B. Kjerfve (Eds.), *Marine ecosystems of Latin America* (pp. 229–244). Springer.

Tarazona, J., & Valle, S. (1998). La diversidad biológica en el mar Peruano. 103–115, *Acta Zoológica Mex Especial*. En: G. Haffter (comp.). La Diversidad Biológica en Iberoamérica Vol. III, 223 p. Xalapa, México.

Tarazona, J., Gutiérrez, D., Paredes, C., & Indacochea, A. (2003). Overview and challenges of marine biodiversity research in Peru. *Gayana (Concepción), 67*, 206–231. https://doi.org/10.4067/s0717-65382003000200009

Tavares Corrêa, C. (2022). El nivel del mar y la erosión de playas en el Perú. In *Desafíos y perspectivas de la situación ambiental en el Perú: en el marco de la conmemoración de los 200 años de vida republicana* (pp. 342–358). Pontificia Universidad Católica del Perú.

Tejada-Pérez, C. A., Villasante, F., Luque-Fernández, C., & Tejada-Begazo, C. L. (2018). Riqueza y distribución vertical de los moluscos del litoral rocoso de la provincia de Islay, Arequipa, al sur del Perú. *Revista Ciencias Marinas y Costeras, 10*, 47. https://doi.org/10.15359/revmar10-1.4

Thiel, M., & Ullrich, N. (2002). Hard rock versus soft bottom: The fauna associated with intertidal mussel beds on hard bottoms along the coast of Chile, and considerations on the functional role of mussel beds. *Helgoland Marine Research, 56*, 21–30. https://doi.org/10.1007/s10152-001-0098-3

UNEP. (2011). *Taking steps toward marine and coastal management*. UNEP.

Valqui, J., Ibañez-Erquiaga, B., Pacheco, A. S., et al. (2021). Changes in rocky intertidal communities after the 2015 and 2017 El Niño events along the Peruvian coast. *Estuarine, Coastal and Shelf Science, 250*, 107142. https://doi.org/10.1016/j.ecss.2020.107142

Chapter 2
The Natural History of Marine Otter (*Lontra felina*)

Gonzalo Medina-Vogel, Carlos Calvo-Mac, Nicole Delgado-Parada, and Gabriela Molina-Maldonado

Abstract The marine otter, *Lontra felina*, is the smallest and most recently evolved marine mammal found primarily in marine habitats. Populations of marine otters have been identified in freshwater and coastal environments. The marine otter is an opportunistic species, as it adapts its diet based on seasonal and prey availability. The reproductive patterns, activity patterns, distribution areas and social behavior of marine otters are still poorly understood. However, it has been estimated that there is variation in population and densities within their geographic range. Historical threats to the conservation of the species include habitat fragmentation, poaching and bycatch. In recent years, new threats have been studied, such as diseases, oil spills, microplastic and mercury pollution. Scientific knowledge on this mustelid is scarce; therefore, it is necessary to carry out exhaustive research to understand their distribution, behavior and conservation needs to generate better conservation strategies along its great latitudinal distribution.

Keywords Chungungo · Chinchimen · Huallaque · Anzumo · Nutria marina

G. Medina-Vogel (✉)
Centro de Investigación para la Sustentabilidad, Facultad Ciencias de la Vida, Universidad Andrés Bello, Santiago, Chile

IUCN Species Survival Commission, Otter Specialist Group, Gland, Switzerland
e-mail: gmedina@unab.cl

C. Calvo-Mac
Centro de Investigación para la Sustentabilidad, Facultad Ciencias de la Vida, Universidad Andrés Bello, Santiago, Chile
e-mail: carlos.calvo@upch.pe

N. Delgado-Parada · G. Molina-Maldonado
Facultad de Ciencias Biológicas, Pontificia Universidad Católica de Chile, Santiago, Chile
e-mail: nadelgado@uc.cl; gvmolina@uc.cl

L. Ayala et al. (eds.), *Marine Otter Conservation*,
https://doi.org/10.1007/978-3-031-53931-2_2

2.1 Species Description

The marine otter belongs to the order Carnivora, family Mustelidae, subfamily Lutrinae and genus *Lontra*. The genus *Lontra* is monophyletic and comprises four species, three of them occurring inside South America (*Lontra felina*, *Lontra longicaudis* and *Lontra provocax*) (Koepfli et al., 2008). The marine otter (*Lontra felina*; Molina, 1782) is the only species of its genus living nearly exclusively in marine habitats and so, is presumably the most recently evolved marine mammal of the world and the smallest of all (Vianna et al., 2010; de Ferran et al., 2022). Recently, populations of the species occupying permanently freshwater environment in the south of Peru have been identified (Ugarte-Núñez, 2021), (Medina-Vogel per. Obs. See Chap. 6); in fact, the species occasionally dabble into coastal wetlands (lagoons and swamps) (Sielfeld, 1983) (Medina-Vogel per. Obs.). Measurements (mm) of marine otters from Chile, report a total length of 910 mm, tail 340 mm, hind foot 97 mm, total weight 4–4.5 kg (Osgood, 1943); body length 660 mm, tail length 390 mm, ear length 20 mm (Housse, 1953); average total length 900 mm (12, 833–1149), for tail length 340 mm (12, 300–362), for hind foot 95 mm (9) and for ear length 15 mm (Redford & Eisenberg, 1992). Average weight 3.6 kg, 3.1–4.5 kg and less than 100 cm total length, measurements of 24 individuals' life captured and released between 2007 and 2019 in Chile within research activities (Medina-Vogel unpublished data). Marine otters are characterized by having a dark brown fur on their back and sides and slightly lighter on the underside. Unlike *L. longicaudis* and *L. provocax*, marine otter shows no sexual dimorphism in size or color, but also has large vibrissae, long tapered tail, and fully webbed feet (Lariviere, 1998) (Fig. 2.1). Juveniles are darker than adults (Redford & Eisenberg, 1992). The species has several names. Spanish: chungungo, gato marino, lobito marino, nutria de mar, nutria marina (in all regions), chinchimén, chingungo, gatuna, chimchimko, s., gato marino que tiene su cueva en las piedras a orillas (Chile), anzumo, zorrillo, pejegato, huallaque, (Peru); English: South American Sea otter, marine otter, sea cat (Valqui, 2012; Pizarro-Neyra, 2022). To differentiate from the other two South American species, *L. felina* has the rhinarium bare and has a straight dorsal border, in contrast to the biconcave edge of the Southern river otter (*L. provocax*) and partially furred rhinarium of the Neotropical river otter (*L. longicaudis*), also these last two are significantly bigger (>6 kg). Furthermore, marine otter skulls never exceed 10 cm in length (Sielfeld, 1983). Marine otter occurs sympatrically with *L. provocax* along southern Chilean Patagonian archipelago from La Araucanía region to Cape Horn (Ebensperger & Botto-Mahan, 1997; Sielfeld & Castilla, 1999). However, there is no data about co-occurrence of both species using same latrines or study sites. Marine otters do not have external distinctive marks to be individualized by direct observation (Lariviere, 1998).

Fig. 2.1 Physical characteristics of *Lontra felina.* The image shows: (**a**) elongated and thick vibrissae, (**b**) long and conical tail, (**c**) fully webbed feet. (**a**) Photograph by Carlos Calvo-Mac, (**b**) Borja Bernales Santolaya, (**c**) Carlos Calvo-Mac

2.2 Historical Records

The marine otter was first described in Chile by the Jesuit priest Juan Ignacio Molina (Molina, 1782). Subsequent publications of the species are scattered (Kerr, 1792; Bennet, 1832; Waterhouse, 1839; Schinz, 1844), described the species as *Lutra chilensis* Gray in 1865 as *Nutria feline* (Gray, 1865). In 1835, during his five-year journey on HMS Beagle, Charles Darwin recorded marine otters at the Chonos Archipelago, Chile (Darwin, 1889). In 1844, the Swiss naturalist Johann Jakob Tschudi, reported the species as very frequent along Peruvian coasts (Tschudi, 1844). Gay in 1874 carried out descriptive studies on the species (Gay, 1874). Thomas in 1908 described *Lutra incarum* (Thomas, 1908), also described by Tschudi in 1844 as *Lutra montana* (Tschudi, 1844), later Pohle in 1920, Thomas in 1920 and Colyer in 1936 as different species from *L. felina* distributed along south of Peru and Bolivia, type locality of Marcapata, Cuzco province, Peru (Pohle, 1920; Thomas, 1920; Colyer, 1936). Probably observations confound with *L. longicaudis*. The coasts of Peru have been inhabited by different pre-Hispanic cultures. The people who have inhabited the Peruvian coasts have had contact with *L. felina* and this has been recorded over time. For example, the word anzumo, from the Yunga language, is currently used in the region of La Libertad, on the northern coast of Peru (Pizarro-Neyra, 2022). In addition, relate the word anzumo with representations of otters and/or squirrels in friezes from the pre-Columbian Chimu of Chan, located in the city of Trujillo in the region of La Libertad (López Austin & Millones, 2008). Also, from the northern coast of Peru, two representations of otters in ceramics are found in the collection of the Larco museum in Lima (Museo Larco, n.d.-a, n.d.-b). Additionally, in the city of Trujillo, Rosello et al. (2001) and Vásquez and Rosales (2016) found remains of *L. felina* in the archaeological zone of Huaca de La Luna (Vásquez & Rosales, 2016); however, the methodologies employed for the identification of these remains as *L. felina* lack clarity, particularly in light of the presence of other mustelids and carnivores in the northern coast of Peru. Similar findings occurred in the Atacama Desert, in Chile, where remains attributed to *L. felina* were identified in an archaeological site (Olguín et al., 2015). Finally, Schaeffer in 2006 identified *L. felina* with the name anzumito in traditional stories of fishers in Huanchaco, a fishing village located north of the city of Trujillo, also in the region of La Libertad, Peru (Schaeffer, 2006). In the other limit of the species' distribution, Cabrera in 1957 reported frequent sightings in scientific expeditions in Tierra del Fuego and Isla de Los Estados in Argentina (Cabrera, 1957). There are records (skulls) of the species in Argentina in Isla de Los Estados (54°49′S; 64°30′W), in the Beagle channel, and in the Mitre Peninsula (54°56′S) (Parera, 2002). However, there are no new observations and records in Argentina (Cassini, 2008). In fact, *L. felina* was described as nearly extinct in Cape Horn (Harris, 1968). Currently is considered data deficient in Argentina (Secretaría de Ambiente y Desarrollo Sustentable de la Nación y Sociedad Argentina para el Estudio de los Mamíferos, 2019). To the north, in 1959, the German naturalist Ernst Schweigger travelled along the Peruvian coast and documented marine otters up to Isla Lobos de Tierra (6°26′S) (Schweigger, 1964). According to Housse in 1953, marine otter distributed along all the Chilean seashore, including observation of the species along the docks of the seaport of Valparaiso city (33°01′S), between Algarrobo and Tunquen (33°19′S). Also observed south of Tomé (36°38′S), Tirúa (38°20′S), in the Ancud Gulf (41°51′S), Llancahé channel (42°03′S), Guaitecas and Chonos Archipelago (43°46′S–46°36′S) (Housse, 1953).

2.3 Distribution and Abundance

There have been several intentions to determine marine otter population densities and abundance along its geographical range, even to count them and compare between local groups. However, the lack of sexual dimorphism, external natural marks, the species elusive behavior, daily (night and day) activity patterns, and spatial behavior (core area and home range overlap) are fundamental obstacles for accurate population estimates and underlined the strong variation of density values obtained in several studies done in past years (Sielfeld, 1992; Medina-Vogel et al., 2007, 2008; Valqui, 2012). Thus, results about estimating population densities (otters/km) by direct observation must be interpreted with caution. Furthermore, most studies assessing the marine otter population only covered limited regional areas and applied different methodologies to determine the total population, making impossible any comparison (Sielfeld & Castilla, 1999; Valqui, 2012). For example, without explaining their methodology Castilla and Bahamondes in 1979 estimated about 200–300 individuals for the Peruvian coast (Castilla & Bahamondes, 1979). Similarly, Vaz-Ferreira in 1979 proposed the population of Peru, Chile and Argentina consisting of 1000 individuals (Vaz-Ferreira, 1979).

By collecting marine otter feces, in two regions and five localities along the coast of Peru separated by ~730 km, in the region of Lima and in the regions of Moquegua and Tacna, covering a total of 18.14 km of coastline, and applying molecular techniques, Biffi & Williams identified a minimum of 80 individuals (41 males and 39 females) across all locations for a density estimate of 4.4 otters/km, after estimates using specialized software (CAPWIRE) they identified an average of 12.6 otters/km (Biffi & Williams, 2017). Indeed, they found a strong positive relationship between the number of fresh scats and the number of unique genotypes, suggesting scat counts might be used to estimate the minimum number of otters at a site. Previously, Apaza et al. (2004) without molecular analysis proposed a population estimate of 691 ± 76 individuals for central and southern Peru (Apaza et al., 2004). That is an average density of 2.21 ± 1.97 individuals per kilometer (otter/km). Mangel and Alfaro-Shigueto (2004) registered 22–28 individuals in 15.6 km of coast between Ilo (17°42′S) and Vila-Vila (18°07′S) in southern Peru (Mangel & Alfaro-Shigueto, 2004). The population in 55 km of rocky shore was estimated at 88 (±11) individuals (Valqui, 2012). In Chile, by using minimum abundance indices, that is counting the maximum numbers of otters active at the same time in a defined area for 1 year by three observations (June 1999 to June 2000), involving a total of 320 hours of observation along three separated localities near Valdivia (39°40′S), the average estimated density was 3.8 otter/km (Medina-Vogel et al., 2006). Furthermore, a single radiotracking study in central Chile (33°11′S) demonstrates a high variation in otter/km as a response to food availability, from 0 to 6 otter/km. Sometimes with 100% home range overlap (Medina-Vogel et al., 2007).

Using the known field methods of direct observation of an otter in the water or on the ground, and the registration of otter field signs (scats, footprints), the first significant survey of the species was done by Walter Sielfeld along seashores of several islands in Magallanes, Southern Chile in the 1980s (Sielfeld, 1989, 1990a, b). Thereafter, the first continental survey was done by Medina-Vogel et al. (2008), between 2004 and 2006 authors studied the occurrence of marine otters along ~2000 km of continental seashore in Chile (Medina-Vogel et al., 2008). Later, Apaza and Romero from referential and field studies described the distribution of *L. felina* along the coast of Peru (Apaza & Romero, 2012). Nowadays, distributional reports are associated to specific localities where research is carried out, and no record has been listed in Argentina and Ecuador (Apaza & Romero, 2012; Valqui, 2012). Indeed, the species register a patchy distribution from Peru to Cape Horn (Chile), associated to a variety of habitat physical characteristics and prey availability (Medina-Vogel et al., 2008; Mangel et al., 2011; Apaza & Romero, 2012; Valqui, 2012). In the Northern limit of Peru the species have been observed only in two occasions, one 115 km to the north of Chimbote, in the artisanal port of Huanchaco

(8°04′S) (Alfaro-Shigueto et al., 2011) and the second in Salaverry (8°13′S) (Santillán & Caro, 2007) (Fig. 2.2). Instead, between Río Santa (08°58′S) and Peninsula Paracas (13°47′S) otters are observed frequently along the coast (Apaza & Romero, 2012), suggesting that to the north of Río Santa there are no permanent

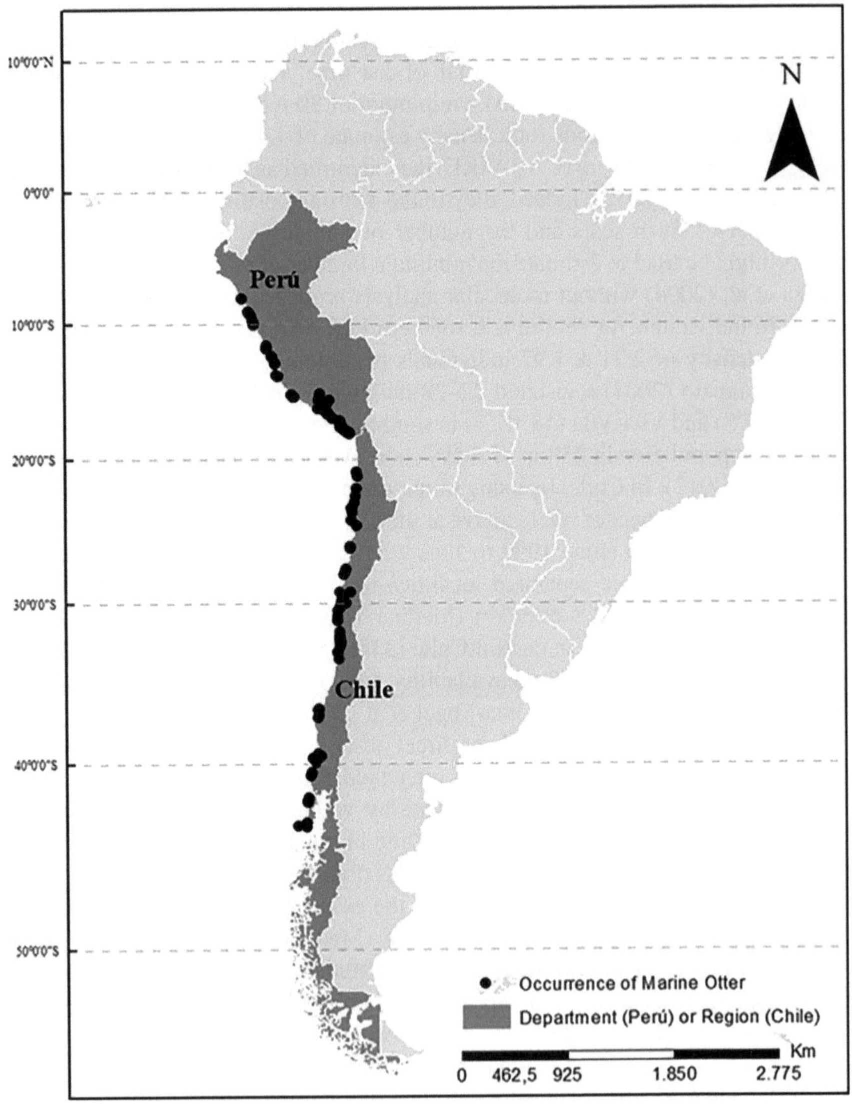

Fig. 2.2 Current distribution of the marine otter (*Lontra felina*) based on data from scientific literature published in the twenty-first century. The dark gray dots represent the occurrence points, which range from Huanchaco (8°S) in the Libertad region of Peru to Guafo Island (43°S) in the Los Lagos region of Chile

populations of marine otter. That northern coastal seashore of Peru is mostly plain and desert, a geomorphological characteristic much less frequent toward the south (Apaza & Romero, 2012). From Río Santa to the border with Chile, the Peruvian coast encompasses ~1500 km of about 100 rocky shore patches offering suitable conditions for marine otters alternated by patches of unsuitable habitat (e.g., sandy beaches and/or shoreline without caves) (Valqui, 2012). Interruptions average 9.1 km in length and are maximal 98.4 km long (Valqui, 2012). Some conspicuous groups of marine otters are present in Pucusana (12°28′S), Punta Corrientes (12°57′S), San Gayán Island (13°50′S), Morro Sama (17°59′S), Vila (18°07′S), Quebrada de Burros (18°10′S) and Boca del Rio (18°90′S) (Sánchez, 1992; Apaza et al., 2004; Valqui, 2004; Ruiz, 2009; Valqui et al., 2010; Mangel et al., 2011; Ortiz-Alvarez et al., 2021) and recently, the population north of the city of Lima, the most populated city in the region, where an oil spill altered the habitat of this population and lead dead marine otters (Hooker & Pizarro-Neyra, 2022). Thus, marine otter occurrence in Peru concentrates between the southern border of the region of Ica and the norther border of the region of Ancash, including the region and city of Lima, becoming scattered distributed from Ica and the border with Chile (Apaza & Romero, 2012). In the region of Arequipa, marine otters have sometimes been observed in the Majes and Cañete rivers, several kilometers away from the ocean (Grimwood, 1968; Apaza et al., 2004), feeding on shrimps in river farms of Camaná and Ocoña up to 650 m above sea level (Hvindberg-Hansen, 1970; Tello, 1972; Brownell, 1978; Viacava Campos et al., 1978). In fact, the observation of one marine otters and latrines were made in the Mamacocha lagoon (15°39′S; 72°16′W) ~1700 m.a.s.l. and more than ~157 km from the sea following the river Colca (G. Medina-Vogel and J. Valqui pers. obs., 2017; see chap. 6).

Between the region of Tacna (South of Peru), the border with Chile and the Chiloe Island (Fig. 2.2), the Chilean rocky seashore is discontinued, with sandy beaches, similarly to the Peruvian coast and it encompasses ~3000 km of areas with marine otter presence alternated by marine otter free areas (Medina-Vogel et al., 2008; Valqui, 2012). Geographically rocky seashore patches are separated increasingly, from northern Chile to central-south Chile, in this region the otter occupancy of rocky seashore is significantly influenced by the presence of humans, in fact marine otter occupancy on surveyed rocky seashore patch was significantly lower in those sections and in rocky patches associated with high levels of human influence of the seashore (Medina-Vogel et al., 2008). Also in this region, the significantly higher marine otter occupancy was recorded only in the largest rocky seashore patches (~412 km long) (Medina-Vogel et al., 2008). As expected, a pattern between increasing distance to the nearest large rocky seashore patch and decreasing marine otter occupancy was not apparent in those study sites with high human influence but was clear in those sections associated with low human influence (Medina-Vogel et al., 2008). Consequently, when analyzing patch size (length) against isolation, it becomes clear that the size and isolation of the rocky seashore network within an area are significantly associated with the occupancy incidence of marine otters. This relationship is altered by the human influence on the seashore (Medina-Vogel et al., 2008). Furthermore, the diet of marine otter becomes more piscivorous towards the north, where occupancy incidence of marine otters is higher (Medina-Vogel et al., 2008; Mangel et al., 2011). Between Algarrobo (33°22′S) and Peninsula Tumbes (36°36′S) and between Arauco Bay (37°10′S) and Queule (39°23′S) no marine otters were reported (Medina-Vogel et al., 2008; Vianna et al., 2010). And conspicuous groups are reported for Mejillones (23°04′S), Constitución (23°54′S), Pan de Azucar (24°40′S), Soldado Bay (26°10′S), Isla Choros (29°15′S), Arrayán (29°43′S), Panul (30°03′S), Las Tacas (30°05′S), Palo Colorado (32°01′S), Puquén (32°11′S), Papudo (32°26′S), Cachagua (32°35′S), Quintay (33°11′S), Hualpén (36°46′S), Tumbes (36°42′S), San Vicente (36°43′S), Chome (36°46′S), Colcura (37°07′S), coast of Valdivia (39°39′S), Pucatrihue (40°10′S), Bahía Mansa (40°34′S), Punhihuil (41°55′S), Chiloe Island (42°10′S), Quilán Island (43°23′), and Guafo Island (43°35′) (Rozzi & Torres-Mura, 1990; Villegas et al., 2007; Córdova et al., 2009; Badilla & George-Nascimento, 2009; Vianna et al., 2010; Cursach et al., 2012; Poblete et al., 2019; Gutiérrez et al., 2019; Calvo-Mac et al., 2020) (Fig. 2.2).

Very little is known about the populations south of Guafo Island and Chiloe Island (43°35′S). Between January and December 1982, Sielfeld studied 11 sites with a total of 67 marine otter dens, seven in the Mornington Island (49°42′S), six in the Contreras Island (52°00′S), two in the Bauclork Island (53°11′S), seven in the Carlo II Island (53°40′S), five in the Skyring (54°23′S), four in the Capitan Aracena Island (54°10′S), four in the Baskot Island, eleven in the Stewart Island (54°52′S), six in the Brocknock channel (54°39′S), eight in the Hosto Island (Seno Monorayo (55°16′S) and seven dens in the Grovy island (Cape Horn) (Sielfeld, 1990a, b). There are no new records for marine otters in Magallanes. Recently, Medina-Vogel et al. (2023) report from anecdotical observations the presence of marine otter in the Johnson Island (44°20′S), Rowlett Island (44°47′S) and the Ninualac sea channel (45°02′S) in the Aysén Region, Chile. Furthermore, there are reports of anecdotical observations of marine otter populations at the Islotes Las Hermanas, Isla Refugio, Palena river estuary area (43°47′S; 73°1′W). These observations are the only recent record south of Chiloe Island. Leaving an open question of how the marine otter populations are distributed between the west coast of Chiloe Island and the Palena river estuary. Thus, marine otter has the longest distribution of all otter species in South America, with extant populations occurring throughout the Pacific coast, all of Chile, and almost all of Peru (Fig. 2.2). Although, its long distribution, marine otter numbers, are in decline due to the presence of environmental pollution, habitat fragmentation, diseases, predation by domestic dogs, and accidental trappings of otters in crab nets (Medina-Vogel et al., 2008; Mangel et al., 2011; Valqui, 2012; Calvo-Mac et al., 2020; Barros et al., 2022).

2.4 Reproduction and Behavior

Several studies describe behavioral aspects, but there are a lot of contradictions concerning reproduction. An adult female collected in February near Peninsula Paracas, Peru, was pregnant (Brownell, 1978). Larivière described the mating season for marine otters to occur between December and January, and parturition to occur from January to March (Lariviere, 1998). Cabello observed matings from December to January, pregnancy March to June, and parturition between May and August (Cabello, 1985). Parera mentioned parturition from January to March (Parera, 2002). Medina-Vogel et al. recorded one lactating female between May and July (Medina-Vogel et al., 2007). Pups and otters mating have been observed in July (Calvo-Mac pers. obs.). However, many authors describe pups mating all year round (Cabello, 1985; Valqui, 2004; Medina-Vogel et al., 2006). It is unknown whether *L. felina* exhibits delayed implantation (Lariviere, 1998). Female marine otters are usually seen swimming with one to two cubs.

A single radiotracking study showed no significant preferences by the species for activity during the day, night, or any other period, in fact otters were 60% active during the night (Medina-Vogel et al., 2007). And 80% of its time otters spend on land, preferably resting in or near their dens (Medina-Vogel et al., 2007). Indeed, during 24 h of monitoring with radiotracking device, marine otters spent significantly more time resting (71%) than moving (17%) or involved in other undetermined activities (Medina-Vogel et al., 2007). Furthermore, of the radiolocations on land, most (87%) corresponded to otters resting, and only 29% (206) radiolocations corresponded to active otters. Of the radiolocations when otters were in the water, 61% were during the night (2010–0800 h) (Medina-Vogel et al., 2007). During this radiotracking study home ranges varied from 1373 m to 4134 m long, core areas varied from 49 m to 495 m long with no significant differences between sexes. Home ranges and the identified core areas followed the seashore contours. Core areas were caves or cavities between rocks or rock cracks used as dens or resting places on land. Core areas averaged 8% of home ranges of males and females. Otters were able to move (maximum distance) between 55% and 151% of their home range extension in less than 24 h (Medina-Vogel et al., 2007). The activity of the six radio-tagged otters (3 males, 3 females) was between 102 m offshore and 30 m on land, although in the same study area marine otter feces were collected up to 50 m on land (Medina-Vogel et al., 2007). On several occasions, otters with or without transmitters were observed simultaneously sharing a group of rock cavities in an area not larger than 100 m^2. However, there were no records of family groups, breeding behaviors or mating, cooperative feeding, or groups of otters moving or resting together in the same den or feeding patches at sea. When there was an artificial offer of food (fishery waste), untagged otters and radio-tracked otters were observed together feeding on waste (Medina-Vogel et al., 2007). Thus, marine otters had small home ranges, females exhibited intrasexual territoriality, and males showed no territoriality. In fact, marine otter appears to show extensive home-range overlap and no avoidances, like other otter species such as *Aonyx capensis*

(Arden-Clarke, 1986), *Lutrogale perspicillata* (Hussain & Choudhury, 1995), *Hydrictis maculicollis* (Perrin & D'inzilio, 1999), and *Lontra canadensis* (Blundell et al., 2002). Thus, it appears that marine otters are solitary, which contrasts with *E. lutris*, a fundamentally group-living otter (Bekoff et al., 1984; Garshelis & Garshelis, 1984), and *L. canadensis*, which in marine conditions forms mixed-sex groups that forage cooperatively (Blundell et al., 2002). Furthermore, marine otters exhibited an opposite spatial behavior to that described for *Lutra lutra* in marine conditions in which females live in groups and males are solitary, rarely overlapping their home ranges and their core areas, which are associated with holes on land and with freshwater pools (Kruuk & Moorhouse, 1991). Although, marine otter is reported as solitary and seldom to be found in groups (Cabello, 1985; Lariviere, 1998; Sielfeld & Castilla, 1999) there are several records of groups of five to eight individuals (Apaza et al., 2004; Valqui, 2004, 2012; Ruiz, 2009; Apaza & Romero, 2012). This nonterritorial behavior of marine otter males and the intrasexual territoriality of *L. felina* females might be a result of the permanent presence of females in estrus during the year, and so, also the presence of pups all year round (Medina-Vogel et al., 2007).

Studies without tagged marine otters do not consider the time marine otters are out of view, so they are likely to be biased in favor of foraging. Indeed, studies about marine otter behavior are done during daylight, representing <10% of their daily movements, interspecific interactions, hunting behavior, resting, breeding and social behavior, so those studies should be interpreted with caution. Nevertheless, several daylight observational studies stated that the marine otter displays activity peaks at specific hours during the day (Ostfeld et al., 1989; Sánchez, 1992; Valqui, 2004; Ruiz, 2009). Long-distance swimming between rocky patches and sandy beaches is eventually used for resting (Ebensperger & Carlos, 1992; Valqui, 2012). Otter nests are built and rebuilt repeatedly after hightide washed them away (Valqui, 2012). To forage, they start to search for their prey once arrived at the bottom. Thus, dive length is correlated to water depth (Medina, 1995). Foraging success ranges from 18% to 32% of register dives (Ostfeld et al., 1989; Medina, 1995; Valqui, 2004). If captured prey are larger than the otters' head, they are carried in the mouth and may be put on belly while swimming on their back and brought to shore. Prey is then consumed with the help of the forepaws (Medina, 1995; Valqui, 2012).

2.5 Adaptation to Marine Conditions

Studies analyzing marine otter diet and prey availability on a monthly and geographical variation concluded that the species showed opportunistic feeding behavior, selecting prey seasonally and locally according to their availability rather than their energy input, and so prey availability affects otter diet and so the presence/absence and density of marine otter populations (Medina-Vogel et al., 2004; Mangel et al., 2011). Unfortunately, most of the studies about the diet of marine otter significantly differ in the number of analyzed scats and prey remains sampled, which make comparison difficulty (Córdova et al., 2009). Nevertheless, comparative studies record a latitudinal pattern regarding prey consumption, towards the southern distribution in Chile, marine otter predates mostly on crustaceans, followed by fish and, to a lower extent, mollusks (Córdova et al., 2009). Thus, marine otter is an important predator of benthic invertebrates and exhibits an opportunistic feeding pattern characterized by a higher benthic prey diversity in the south-central area of the Chilean coast (Córdova et al., 2009). Instead towards the north of Chile, marine otter predates mostly on fish followed by crustaceans (Mangel et al., 2011). In the region of Tacna (18°00′S), Peru, a study involving the analysis of 66 marine otter scats collected in two study sites Morro Sama (MS; 17°59′42.8″S) and Quebrada Burros (18°01′42″S), described a diet composed of 22 different species of prey, 59.1% (relative frequency) arthropods ($n = 13$), 27.3% fish ($n = 6$) and 13.6% mollusks ($n = 2$) with seasonal variation (Biffi & Iannacone, 2010). In the same area, identify prey of marine otter by direct observation were mostly fish (41–48%) followed by crabs (20–29%), shrimps (1–15%) and other unidentified preys (Mangel et al., 2011). The largest diet study in Chile, involve the analysis of 475 scats collected in four different study sites close to Valdivia (39°40′S), from which 78.4% (relative frequency) corresponded to thirteen crab species, followed by 20.2% ten fish species and 1.4% two species of mollusks (Medina-Vogel et al., 2004). Another recent study in the south of Chile, described a low diet diversity, from a total of 131 scats collected (65 in Caleta Chome 36°46′S and 66 in San Vicente; 36°44′S), these consisted mainly of crustaceans (75–88%), followed by fish (13–21%) and shellfish (3%) (Poblete et al., 2019). In Magallanes, their southernmost limit of distribution, marine otter recorded a uniformly diet composed of fish and crabs (Sielfeld, 1990b). Marine otter diet and feeding behavior are also affected by weather and sea condition. Indeed, Villegas et al. (2007) studied the daylight foraging activity of marine otter and compared results between wave-protected site and wave-exposed sites (Villegas et al., 2007). They concluded that marine otters spent more time foraging in the wave-protected site compared with the wave-exposed habitat. Successful dives reached 26.9% in the wave-exposed habitats, and 38.2% in the wave-protected habitat. Furthermore, they recorded dives 18% shorter in wave-exposed as compared with wave-protected habitat. Similarly, Gutiérrez et al. (2019) register significantly more time dedicated to displacement of marine otter in Quintay (3°11′S) during summer, no difference during fall, and significantly more time dedicated to dive in Quintay in both seasons, compared to a northern site Cachagua (32°35′S) (Gutiérrez et al., 2019). But they recorded no significant differences in feeding behavior between seasons or localities. The major time spent otters diving in Quintay than in Cachagua in both seasons, is consistent with the conclusion that wave-protected habitat is better for marine otter (Villegas et al., 2007). Marine otter also predates birds (Sielfeld, 1990b; Medina, 1995). Indeed, during fieldwork on Isla Choros (29°15′S) marine otters were registered actively preying upon at least two Peruvian Diving Petrel chicks (*Pelecanoides garnotii*) (Mattern et al., 2002). Urchins were found in only 2.9% of the scat samples but were over half of the prey remains found in dens in Magallanes and therefore seem to be an important part of the diet (Sielfeld, 1990b). Urchins have been occasionally observed only in other latitudes. Marine otters preying on octopuses have been observed in Peru and Chile (Medina, 1995; Valqui, 2004, 2012; Medina-Vogel et al., 2004).

The marine otter is the most recent mammal adapted to marine life (Estes, 1989). The other is the sea otter (*Enhydra lutris*), a larger species (10–40 kg body weight) evolved a high metabolism to feed frequently and abundantly to maintain its body heat in the cold waters of the Pacific Ocean in the Northern Hemisphere and the Bering Sea (Morrison et al., 1974; Bowen et al., 2002), and the densest and most waterproof fur of all aquatic mammals. Sea otter spends at least 12% of its time per day is spent grooming to maintain the waterproof and insulating properties of its fur (Williams & Worthy, 2002). However, marine otter is the smallest of its genus (Estes, 1989). Contrary to the sea otter, the marine otter spends most of its daily time out of the water resting inside caves or rock cracks (Medina-Vogel et al., 2006, 2007). The little time that marine otter spends in the water may be due to an evolutionary adaptation or a behavioral response to an environment with abundant prey close to shore, where dives into the water are short and highly successful (Ostfeld et al., 1989; Medina, 1995; Valqui, 2004; Medina-Vogel et al., 2004). Contrary to the sea otter, the evolutionary adaptation of the small marine otter might well be a reduced time in the water. An adaptation is only possible in an environment where dives searching for prey most of the time are successful. Indeed, marine otters are well adapted to aquatic conditions due to their powerful tail and webbed feet (Lariviere, 1998). And due to their small size and reduced body fat, heat loss is reduced by avoiding long-time periods in the water. Marine otters have been observed swimming 2 km in about 20 min (Valqui, 2012), forage in shores with strong waves (Cabello, 1985; Ostfeld et al., 1989; Villegas et al., 2007) and dive between 5 s and 77 s with no relationship between dive length and hunting success, or between the dive length and prey size (Medina, 1995; Valqui, 2004). They even spend time swimming and feeding in the short period of time between two breaking waves (Valqui, 2012). However, most river otters are shallow-water divers. In fact, Eurasian otters (*Lutra lutra*) have limited function according to water temperature, and hence their time spent diving is reduced in cold water. For Eurasian otter the costs of diving in water at 2 °C are some 2.7 times higher than they at 20 °C (Kruuk et al., 1994). So for marine otter, a semi-aquatic species inhabiting a coast washed by the cold Humboldt current coming from the southern Pacific and the Antarctic, the sea conditions, the depth and time needed to spend in the water searching for pray must be a limiting environmental factor, and might explain the ~80% of their daily time spent inland resting, and the strong negative effect that human overexploitation of subtidal and intertidal invertebrates and fish could have on marine otter distribution and population densities.

Marine mammals are hypotonic, sea otters create urine in elevated concentrations (Costa, 1982). They regulate their salt balance by hormonal regulation of urine concentration and/or by the rate of urine concentration (Ortiz, 2001). This osmotic ability allows sea otters to consume sea water and gain free water by excreting excessive Na^+ and Cl^- (Ortiz, 2001). Marine otter living close to rivers might directly drink freshwater. In fact, several times marine otters have been observed drinking freshwater from streams, even from hose pipes near the sea along the coast, and in captivity they frequently drink freshwater (Javier Tribelli and Medina-Vogel pers. obs.). Thus, it is unknown how the scarcity of access to freshwater affects marine otter occurrence along the coast of central north, north of Chile and along the coast of Peru. And what environmental variables limit the occurrence of marine otter populations in freshwater habitat others than those populations of the South of Peru.

2.6 Phylogeography and Population Genetic

Within the genus *Lontra*, the split time between the North American river otter (*Lontra canadensis*) and the three Neotropical species (*L. longicaudis*, *L. felina*, and *L. provocax*) is estimated in 3.7 mya. Coincides with the formation of the Panama Isthmus ~3 mya, supporting the view that the three species diversified during and after the Great American Biotic Interchange (de Ferran et al., 2022). However, marine otter diverged only 0.43 mya, possibly via parapatric speciation, and reproductive isolation from southern river otter (*Lontra provocax*), whose distributions overlap in southern Chile (de Ferran et al., 2022). Furthermore, de Ferran et al. (2022) investigated genome-wide historical trends in effective population size (N_e) among 11 otter species. The most extreme declines were observed for the two southernmost South American species (*L. provocax* and *L. felina*), likely driven by habitat changes (e.g., extensive ice sheet coverage) during the Last Glacial Period that may have severely affected their population numbers, particularly given that marine otter has a linear distribution restricted to rocky shores along the western coastline of South America.

Previous study carried out in Chile and Peru sequencing the control region of mtDNA in 168 out of 419 collected samples of blood (16), tissue (46) and scat (357). Analysis of 2261 bp of mtDNA (control region, ND5 and Cyt-b sequences) yielded 26 mtDNA haplotypes and microsatellite genotyping, 89 individuals were identified within the 126 scat samples processed (Vianna et al., 2010). They detected a significant population sub-structure and the overall variability values were $h = 0.93$ and 0.0047. Those results are in the range of the study done in Peru by Valqui et al. (2010). Furthermore, Vianna et al. (2010) determined six groups (based on a spatial analysis of molecular variance, SAMOVA), with the Peruvian group separating from Chilean groups, while the two southernmost groups showed reduced genetic diversity (Chile Central-South: $h = 0$ and 0; Chile South: $h = 0.39$ and 0.00069) as it is often recorded in peripheral populations (Vianna et al., 2010; Valqui et al., 2010). Valqui et al. (2010) completed a study along the entire distribution of marine otter in Peru, 87 scats were collected and sequenced mtDNA, only in 41(47%). Valqui et al. (2010) estimated 24 individuals with 11 haplotypes and 13 polymorphic sites. Contrary to Vianna et al. (2010) in Chile, no population substructure (based on F-statistics) and no isolation by distance (based on a Mantel-test) were detected. Considering the low population numbers reported, Valqui et al. (2010) study yielded surprisingly high values of haplotype diversity (h) = 0.86 and nucleotide diversity (π) = 0.0017. Based on geographic criteria, there are four separated group along the Peruvian coast (Valqui et al., 2010). All three genetic studies on marine otter have not yet detected if the species has gone through a bottleneck. In fact, studies of Valqui et al. (2010) and Vianna et al. (2010) suggest gene flow along the species distribution. Indeed, Biffi and Williams (2017) after collecting 240 samples of marine otter feces from seven localities in Peru during four visits, managed to successfully amplify five to seven microsatellite loci and a sex-linked marker in 133 samples (55%). Within that larger sample, they identified a minimum of 80 individuals (41 males and 39 females) across all locations for a density estimate of 4.4 otters per km. Southern river otter is capable of dispersing ~50 km following river courses (Medina-Vogel, 2005). A somehow similar behavior following seashore contour might cause remixing of genetic material in the marine otter population. However, there is an increasing habitat fragmentation associated to human use of the seashore, which could affect marine otter dispersion and increase genetic isolation between populations separated by great distances between patches of rocky shore (Medina-Vogel et al., 2008; Vianna et al., 2010).

2.7 Health and Handling

The combination of ketamine with medetomidine followed by antagonism with atipamezole is a good short-term anesthetic protocol for marine otters (Soto-Azat et al., 2006). Of the nine tested otters, rectal temperature decreased for most otters during the anesthetic period. Six marine otters developed mild hypothermia (<36.7 °C). Authors emphasized the need of using external supplementary heating during anesthesia and recovery even inside a warm room. Hyperthermia (>40.1 °C) was recorded in only one animal (43.5 °C) in the first 5 min of anesthesia. In that study, the mean cardiac rate was 132 ± 20 beats/min and remained generally stable throughout the monitoring period, although a mild to moderate sinus arrhythmia was observed in six otters. Mean arterial blood pressure measured in four otters was 115 ± 30 (range: 64–163) mm Hg during the 30 min of monitoring (Soto-Azat et al., 2006). The respiratory rate ranged from 8 to 44 breaths/min for most monitored otters. However, during monitoring one otter became extremely tachypneic (88 breaths/min) and another otter became apneic within 5 min of anesthetic administration. The anesthetic effects of ketamine and medetomidine in marine otters were like those produced in North American river otters (*Lontra canadensis*) and Eurasian otters (*Lutra lutra*) at similar dosages (ketamine, 5 mg/kg; medetomidine, 50 mg/kg) and in Asian small, clawed otters (*Aonyx cinereus*) at higher dosages of medetomidine (120 μg/kg) with the same dosage of ketamine (5 mg/kg) (Lewis, 1991; Spelman et al., 1994; Fernandez-Moran et al., 2001; Soto-Azat et al., 2006). In Chile there have been already three studies involving the life trapping of marine otters, the three studies use Victor leg-hold traps (1.5 soft catch; Woodstream Corp., Lititz, Pennsylvania) but modifying springs to 0.8–1.0 (Medina-Vogel et al., 2007;

Calvo-Mac et al., 2020). Furthermore, floating traps have been tested with marine otter with success, but they can be used only in specific sites protected from seaweaves (Fig. 2.3) (Medina-Vogel per. Obs.). Traps were checked every 6 h and 12 h respectively. In these two studies otters were anesthetized with a combination of ketamine (5.3 mg/kg) and medetomidine (53 ug/kg) and reversed after 45 min with atipamezole (226 ug/kg) as described by Soto-Azat et al. (2006). One single successful radiotracking study done in Chile use implanted flat radio-transmitters with an activity sensor (150.00–151.00 Mhz; Sirtrack Ltd., Havelock North, New Zealand) in their abdominal cavities. Radio-transmitters were flat, square and 0.4% of the otter's body mass (Medina-Vogel et al., 2007). During transport and release procedures the otters were placed in a well-ventilated transport tube constructed from a 1 m section of 40 cm diameter PVC pipe (Serfass et al., 1996) (Fig. 2.4), for short periods of handling a plastic mesh dubbed while the otter is inside it is recommended (Fig. 2.5) (Calvo-Mac & Medina-Vogel pers. obs.). For all otter handling and captivity, special consideration must be taken to avoid contact with domestic cats and dogs. In North America, Canine Distemper Virus (CDV) has been reported in American mink (*Neogale vison*) and North American river otters (*Lontra canadensis*), also Europe Eurasian otters (*Lutra lutra*) have tested seropositive

Fig. 2.3 Use of floating traps to capture marine otters (*Lontra felina*). The traps are made of a sturdy metal frame and netting and are designed to float on the surface of calm waters. The top of the cage is open, allowing the otters to swim in and out freely, but once inside, they cannot escape. This method of capture is only feasible in areas with calm waters, as rough waves could damage or overturn the cages. (Photo by René Monsalve Alarcón and Vanessa Ilukewitsch)

Fig. 2.4 Transport and release procedures for marine otters (*Lontra felina*) using a well-ventilated transport tube made from a 1-meter section of 40 cm diameter PVC pipe. This method is an adaptation of Serfass et al., 1996 methodology and provides a safe and efficient way to transport marine otters while ensuring they have adequate ventilation. The tube allows for easy release of the otters into their natural habitat, minimizing stress on the animals during the transport process. (Photo by Rene Monsalve Alarcon and Vanessa Ilukewitsch)

Fig. 2.5 Transport and release procedures for marine otters (*Lontra felina*) using a plastic mesh. This method provides a safe and efficient way to transport marine otters for short periods. (Photo by Carlos Calvo-Mac and Rene Monsalve Alarcon)

against CPV-2 and mink parvovirus (Kimber et al., 2000); new antigenic types of CPV-2 have been isolated from stone martens (*Martes foina*) and other carnivores (Steinel et al., 2001). In Chile Southern river otter (*L. provocax*) tested seropositive against Canine Distemper Virus and Parvovirus (Barros et al., 2022). Furthermore, in south of Chile American mink tested seropositive against Canine Distemper Virus and Parvovirus, dog sample in the same study showed 99% identity with Canine Parvovirus, 99% identity with Feline Panleukopenia, and 99% identity with Mink Enteritis virus. An alignment between these sequences showed 100% identity between minks and 99% identity between mink and dog samples. Therefore, it is expected that disease transmission of both Canine Parvovirus and CDV between domestic dog, cats and marine otters could happen in Peru and Chile (Sepúlveda et al., 2014; Barros et al., 2022). Marine otter also has been tested positive to *Toxoplasma gondii* a worldwide distributed protozoon that infects warm-blooded animals. In fact, *T. gondii* exposure in one marine otter (*Lontra felina*) along the northern and central coast of Chile was register (Calvo-Mac et al., 2020), indicating a low seroprevalence of 5% (1/19) in seashore associated to arid environment. Furthermore, as has been registered in Southern river otter, marine otter can also be vulnerable to *Leptospira* spp. (Calvo-Mac et al., 2024; Medina-Vogel, 2010). In the central coast of Peru, the presence of bacteria with zoonotic potential has been identified in the feces of marine otters. The study revealed the presence of *Vibrio parahaemolyticus*, *Vibrio* spp., and *Aeromonas* sp. Additionally, some species of the Vibrio genus were found to be resistant to antibiotics like ampicillin, and a sample was found to be multiresistant to five antibiotics (Calvo-Mac, 2014). Another study also detected the presence of enteroparasites in the feces of marine otters, including *Eimeria* sp., *Strongyloides* sp., *Anisakis simplex*, and eggs of Ancylostomidae and Metastrongyloidae (Vega-Feria et al., 2018). Recently, a diverse type of microplastics were found in marine otter scats collected in Punta Corrientes (12°57′S), located in the central coast of Peru (Santillán et al., 2020). And marine otters are also affected by pollutants like mercury (Medina-Vogel obs. Pers.).

2.8 Conservation Threats

Diseases: Between 2009 and 2023, 130 otters were registered and handled by SERNAPESCA, Chile, 34 alive, 96 dead. The number of otters recorded dead or alive is increasing since 2009. And an increased tendency of otter dead by indeterminate cause were registered towards 2023. Most of the animals were found in the proximities of crowded cities and towns (98%; see Chap. 8). And most of them on sandy beaches, form inside the motors of a vehicle, inside buildings, or inside the refrigeration system of thermoelectric centrale (Correa & Pizarro, 2023). However, this last increase of marine otter death coincides with the High Pathogenic Avian Influenza A (H5N1) outbreak along the south Pacific region. In fact, all otters registered during 2023 were found dead. Furthermore, identified diseases include toxoplasmosis, leptospirosis, canine adenovirus, canine distemper virus and feline influenza (Medina-Vogel, 2010; Calvo-Mac et al., 2020; Barros et al., 2022). Correa and Pizarro (2023) reported four cases of otter wounded and possibly killed by domestic dogs. Interaction, like otters with dogs and cats in close proximity or using same waste feeding spots from human activity, has been observed in Chile and in Peru (Medina-Vogel et al., 2007; Pizarro-Neyra, 2008; Mangel et al., 2011).

Habitat lost and fragmentation: Habitat use by marine otter is selective and most probably respond to the changes caused by humans (Medina-Vogel et al., 2007, 2008; Cursach et al., 2012; Gutiérrez et al., 2019). Indeed, as explained before there is a strong negative association between marine otter occupancy and human presence, responding to a pattern between increasing distance to the nearest large rocky seashore patch and decreasing marine otter occupancy when more human presence and activity is recorded.

Fisheries interactions: Although marine otter preys on resources exploited by artisanal fishing, without damaged their fishing gears. Indeed, a study done in Huenteyao (40°28′S; 73°43′W), report that 66.7% of the fishermen said that otters do not affect their fishing, while 30.0% said the opposite; holding as cause the consumption by marine otter on the same resources exploited by them (e.g., *Concholepas concholepas*) (Córdova & Rau, 2016). Indeed, there is a synanthropic behavior of marine otter along its geographic distribution. The marine otter uses fishing ports infrastructure for shelter and food. It forages on the remains of fish discarded by fishermen which affect the otters' spatial and social behavior (Medina-Vogel et al., 2007; Cursach et al., 2012). Correa and Pizarro (2023) recently report three cases of otter death as a result of being caught and drowned by fishing and crab gears, three cases of otter death presumably as a result of being run over by boats. Furthermore, when fishery waste from fishing landings is available, marine otters have been observed together feeding on waste and sometimes being persecuted by street dogs (Medina-Vogel et al., 2007). Moreover, and possibly as part of the effect of human waste on the otter diet, in fact, diverse type of microplastics were found in marine otter scats collected in Punta Corrientes (12°57′S), located in the central coast of Peru (Santillán et al., 2020). And recently, interaction between marine otter and salmon farming infrastructure have been documented but undescribed south of Chiloe Island (Chile) (Medina-Vogel et al., 2023).

2.9 Research Recommendations

As presented above, scientific information about the marine otter is scarce, geographically dispersed, with almost no information between Chiloe Island and Magallanes (Chile) and in the northern distribution (Ancash and La Libertad region in Peru). Furthermore, studies have concentrated on diet and behavior during daylight, so basic knowledge is needed to understand the current state of marine otter populations along their entire distribution, especially south of Chiloe Island (Chile) and carry out conservation actions accordingly. The more pressing issue is establishing an occurrence map of the species along Peru and Chile, including those populations occurring in freshwater habitats, and how this occurrence differs according to the rocky seashore patches length and spatial distribution, island size and geographical distribution (south of Chiloe Island), and human occupation (residential, mining, energy generation industry, aquaculture, fishing, seaweed forests exploitation). This can be done by setting up a collaborative network to collect new data nationwide, for example using online questionaries and web resources. Furthermore, monitoring programs should be used by combining methodologies to ascertain marine otter presence, such as detection of scats, hair trapping, genetic species identification of scats, and camera-trapping campaigns. An example of successful large extension effort to evaluate marine otter distribution and patterns of occurrence by scats and live detection is the survey done by Medina-Vogel et al. (2008), methodology later supported by genetic identification in scats and tissue samples (Vianna et al., 2010). Another example to assess population density and distribution by genetic identification in scats is those in Peru by Valqui et al. (2010) and Biffi and Williams (2017). Furthermore, there is a need for standardization of visual estimation census. In this regards, Medina-Vogel et al. (2006) methodology proved to be trustworthy, as come to similar conclusions with the later study applying radiotracking and animal tagging of Medina-Vogel et al. (2007), marine otters spend more than 80% of their time inland. Once a distribution baseline is established, scats and hair trapping can be employed to assess dispersion and/or isolation between marine otter population, depending on the rocky seashore patches distribution and human use of the seashores (Vianna et al., 2010). Into this respect, further studies to assess how habitat modified by humans and artificial habitats (shipwrecks, docks, etc.) are used by marine otter as steppingstones between rocky seashore patches widely separated from each other are needed, especially along central south of Chile and north of Peru.

Future research on the marine otter should also aim to determine physiological limitations (osmoregulation) of the species to the sea conditions and prey availability, local abundance, sex ratio, dispersal rate, and expand incipient knowledge on the use of time and space by the species in a 24 h cycle. Also, Vianna et al. (2010) recommended further investigations on male dispersal patterns using biparental or paternal markers to better understand dispersal and the effective role of habitat structure. Additionally, there is increasing importance of further studies on disease infection from domestic animals and alien introduced American mink (*Neogale vison*), pathogen pollution (Calvo-Mac et al., 2020; Barros et al., 2022), and the ecology of pollution in general, e.g., how biomagnification affects marine otter as a top predator of south Pacific seashore.

References

Alfaro-Shigueto, J., Valqui, J., & Mangel, J. C. (2011). Nuevo registro de la nutria marina Lontra felina (Molina, 1782) al norte de su distribuci6n actual. *Ecological Applications, 10*, 87–91.

Apaza, M., & Romero, L. (2012). Distribución y observaciones sobre la población de la nutria marina Lontra felina (Molina 1782) en el Peru. *Revista Peruana de Biología, 19*, 285–298. https://doi.org/10.15381/rpb.v19i3.1064

Apaza, M., Valqui, J., Mangel, J., et al. (2004). *Estado de Conservación de Lontra felina (Molina 1782) en la Costa Peruana*. Lima.

Arden-Clarke, C. H. G. (1986). Population density, home range size and spatial organization of the Cape clawless otter, Aonyx capensis, in a marine habitat. *Journal of Zoology, 209*, 201–211. https://doi.org/10.1111/J.1469-7998.1986.TB03576.X

Badilla, M., & George-Nascimento, M. (2009). Conducta diurna del chungungo Lontra felina (Molina, 1782) en dos localidades de la costa de Talcahuano, Chile: ¿efectos de la exposición al oleaje y de las actividades humanas? *Revista de Biología Marina y Oceanografía, 44*, 409–415. https://doi.org/10.4067/s0718-19572009000200014

Barros, M., Pons, D. J., Moreno, A., et al. (2022). Domestic dog and alien North American mink as reservoirs of infectious diseases in the endangered Southern river otter. *Australian Journal of Veterinary Science, 54*, 65–75. https://doi.org/10.4067/S0719-81322022000200065

Bekoff, M., Daniels, T. J., & Gittleman, J. L. (1984). Life history patterns and the comparative social ecology of carnivores. *Annual Review of Ecology and Systematics, 15*, 191–232. https://doi.org/10.1146/ANNUREV.ES.15.110184.001203

Bennet, E. T. (1832). *Characters of a new species of otter (Lutra, ErxL), and of a new species of mouse (Mus, L.), collected in Chili by Mr. Cuming*. Proceedings of the Committee of Science and Correspondence of the Zoological Society of London, Part 2. Richard Taylor.

Biffi, D., & Iannacone, J. (2010). Variabilidad trófica de Lontra feline (Molina 1782) (Carnivora: Mustelidae) en dos poblaciones de Tacna (Peru) entre Agosto y Diciembre de 2006. *Mastozoología Neotropical, 17*, 11–17.

Biffi, D., & Williams, D. A. (2017). Use of non-invasive techniques to determine population size of the marine otter in two regions of Peru. *Mamm Biol – Zeitschrift für Säugetierkd, 84*, 12–19. https://doi.org/10.1016/J.MAMBIO.2016.12.006

Blundell, G. M., Ben-David, M., & Bowyer, R. T. (2002). Sociality in river otters: Cooperative foraging or reproductive strategies? *Behavioral Ecology, 13*, 134–141. https://doi.org/10.1093/beheco/13.1.134

Bowen, W. D., Read, A. J., & Estes, J. (2002). Feeding ecology. In A. R. Hoelzel (Ed.), *Marine mammal biology: An evolutionary approach* (pp. 217–246).

Brownell, R. L. (1978). Ecology and conservation of the marine otter (Lutra felina). In N. Duplaix (Ed.), *Proceedings of the first working meeting of the IUCN/SSC otter specialist group* (pp. 104–106). Paramaribo, Suriname.

Cabello, C. C. (1985). *Informe final Proyecto 1409 WWF/IUCN Nutria Marina, Chiloe, Chile*. Ancud.

Cabrera, A. (1957). Catálogo de los mamíferos de América del Sur Revista del Museo Argentino de Ciencias Naturales Bernardino Rivadavia. *Cienc Zool, 4*, 308.

Calvo-Mac, C. (2014). *Presencia de bacterias con potencial zoonótico y resistencia antibiótica en heces de nutrias marinas (Lontra felina) en la costa centro del Peru*. Universidad Peruana Cayetano Heredia.

Calvo-Mac, C., Gutleb, A. C., Contal, S., et al. (2020). Exposure to *toxoplasma gondii* in marine otters (Lontra felina) and domestic cats (Felis catus) in an arid environment in Chile. *Journal of Wildlife Diseases, 56*, 962–964. https://doi.org/10.7589/2019-10-269

Calvo-Mac, C., Martínez-Leiva, G. K., Salgado, M., & Medina-Vogel, G. (2024). Exposure to pathogenic Leptospira and *toxoplasma gondii* in endangered native otters of the Valdivian temperate rainforest ecoregion in Chile. *Aquatic Mammals, 50*. https://doi.org/10.1578/AM.50.1.2024.39

Cassini, M. H. (2008). Present status of Lontra felina in Argentina. *Endangered Species Update, 25*, 57–60.

Castilla, J. C., & Bahamondes, I. (1979). Observaciones conductuales y ecológicas sobre Lutra felina (Molina) 1782 (Carnivora: Mustelidae) en las zonas central y centro-norte de Chile. *Archives Biological Medicine Experiment, 12*, 119–132.

Colyer, F. (1936). *Variations and diseases of the teeth of animals*. John Bale, Sons and Danielsson.

Córdova, O., & Rau, J. R. (2016). Interacción entre la pesca artesanal y el depredador de alto nivel trófico Lontra felina en Chile. *Revista de Biología Marina y Oceanografía, 51*, 621–627. https://doi.org/10.4067/S0718-19572016000300013

Córdova, O., Rau, J. R., Suazo, C. G., & Arriagada, A. (2009). Comparative study of the feeding ecology of the top predator Lontra felina (Molina, 1782)(Carnivore: Mustelidae) in Chile. *Revista de Biología Marina y Oceanografía, 44*, 429–438.

Correa, R., & Pizarro, J. (2023). Mortality of *Lontra fe*lina (Molina, 1782) in Chile (2009–2022) based on reports. *IUCN Otter Specialist Group Bulletin, 40*(3), 145–151.

Costa, D. P. (1982). Energy, nitrogen, and electrolyte flux and sea water drinking in the sea otter Enhydra Lutris. *Physiological Zoology, 55*, 35–44. https://doi.org/10.1086/PHYSZOOL.55.1.30158441

Cursach, J. A., Rau, J. R., Ther, F., et al. (2012). Sinantropía y conservación marina: El caso del chungungo lontra felina en el sur de Chile. *Revista de Biología Marina y Oceanografía, 47*, 593–597. https://doi.org/10.4067/S0718-19572012000300022

Darwin, C. (1889). *Viaje de un naturalista alrededor del Mundo*. Tomo II.

de Ferran, V., Figueiró, H. V., de Jesus, T. F., et al. (2022). Phylogenomics of the world's otters. *Current Biology, 32*, 3650–3658.e4. https://doi.org/10.1016/J.CUB.2022.06.036

Ebensperger, L. A., & Botto-Mahan, C. (1997). Use of habitat, size of prey, and food-niche relationships of two sympatric otters in southernmost Chile. *Journal of Mammalogy, 78*, 222–227. https://doi.org/10.2307/1382655

Ebensperger, L. A., & Carlos, J. C. (1992). Selección de hábitat en tierra por la nutria marina, Lutra felina, en Isla Pan de Azúcar, Chile. *Revista Chilena de Historia Natural, 65*, 429–434.

Estes, J. A. (1989). Adaptations for aquatic living by carnivores. In J. L. Gittleman (Ed.), *Carnivore behavior, ecology, and evolution* (pp. 242–282). Springer.

Fernandez-Moran, J., Perez, E., & Sanmartin, M., et al. (2001). Reversible immobilization of Eurasian otters with a combination of ketamine and medetomidine. meridian.allenpress.com, *37*:561–565.

Garshelis, D. L., & Garshelis, J. A. (1984). Movements and management of sea otters in Alaska. *Journal of Wildlife Management, 48*, 665–678. https://doi.org/10.2307/3801414

Gay, C. (1874). *Historia Física y Política de Chile*. Zoología Tomo I. Maulde y Renou.

Gray, J. E. (1865). Revision of the genera and species of Mustelidae contained in the British Museum. *Proceedings of the Zoological Society of London, 33*, 100–154. https://doi.org/10.1111/j.1469-7998.1865.tb02315.x

Grimwood, I. (1968). Endangered mammals in Peru. *Oryx, 9*, 411–421. https://doi.org/10.1017/s0030605300007328

Gutiérrez, L., Vargas, F., Pinto, P., et al. (2019). Impact of human activities in habitat use and activity patterns of the marine otter (Lontra felina) in Central Chile. *Latin American Journal of Aquatic Research, 47*, 122–128. https://doi.org/10.3856/VOL47-ISSUE1-FULLTEXT-13

Harris, C. J. (1968). *Otters : A study of the recent Lutrinae*. Weidenfeld and Nicolson.

Hooker, Y., & Pizarro-Neyra, J. (2022). *Evaluación rápida del estado de las poblaciones de nutria marina Lontra felina en el sector de mayor afectación por el derrame de petróleo de La Pampilla*. Lima.

Housse, R. (1953). *Animales Salvajes de Chile*. Universidad de Santiago.

Hussain, S., & Choudhury, B. (1995). Seasonal movement, home range and habitat utilization by smooth coated otter in National Chambal Sanctuary, India. *Habitat, 11*, 45–55.

Hvindberg-Hansen, H. (1970). *A survey of the coast otter (Lutra felina) Molina in the Peruvian rivers of Camana and Ocona*.

Kerr, R. (1792). *The animal kingdom, or zoological system of the celebrated Sir Charles Linnaeus; Class I. Mammalia; containing a complete systematic description, arrangement, and nomenclature, of all the known species and varieties of the Mammalia, or animals which give*. Murray & R. Faulder.

Kimber, K. R., Kollias, G. V., & Dubovi, E. J. (2000). Serologic survey of selected viral agents in recently captured wild north american river otters (lontra canadensis). *Journal of Zoo and Wildlife Medicine, 31*, 168–175.
Koepfli, K. P., Deere, K. A., Slater, G. J., et al. (2008). Multigene phylogeny of the Mustelidae: Resolving relationships, tempo and biogeographic history of a mammalian adaptive radiation. *BMC Biology, 6*, 1–22. https://doi.org/10.1186/1741-7007-6-10/FIGURES/5
Kruuk, H., & Moorhouse, A. (1991). The spatial organization of otters (Lutra lutra) in Shetland. *Journal of Zoology, 224*, 41–57. https://doi.org/10.1111/J.1469-7998.1991.TB04787.X
Kruuk, H., Balharry, E., & Taylor, P. T. (1994). Oxygen consumption of the Eurasian otter Lutra lutra in relation to water temperature. *Physiological Zoology, 67*, 1174–1185. https://doi.org/10.1086/physzool.67.5.30163888
Lariviere, S. (1998). Lontra felina. *Mammalian Biology, 871*(575), 1–5. https://doi.org/10.2307/3504377/2600421
Lewis, J. C. M. (1991). Reversible immobilisation of Asian small-clawed otters with medetomidine and ketamine. *The Veterinary Record, 128*, 86–87. https://doi.org/10.1136/VR.128.4.86-A
López Austin, A., & Millones, L. (2008). Dioses del norte, dioses del sur: religiones y cosmovisión en Mesoamérica y los Andes. *Año, 17*.
Mangel, J. C., & Alfaro-Shigueto, J. (2004). Comunidades pesqueras y la conservacion de la nutria (Lontra felina) en el sur de Peru. *Revista de Conservación Regional, 2*, 5–10.
Mangel, J. C., Whitty, T., Medina-Vogel, G., et al. (2011). Latitudinal variation in diet and patterns of human interaction in the marine otter. *Marine Mammal Science, 27*, 14–25. https://doi.org/10.1111/j.1748-7692.2010.00414.x
Mattern, T., Ellenberg, U., & Luna-Jorquera, G. (2002). A South American marine otter Lontra felina preys upon chicks of the Peruvian diving petrel Pelecanoides garnotii. *Marine Ornithology, 30*, 95–96.
Medina, G. (1995). Feeding habits of the marine otter (Lutra felina) in southern Chile. *Proceedings of the International Otter Colloquium, 6*, 65–68.
Medina-Vogel, G. (2005). Estrategia regional para la conservación del huillin (Lontra provocax) en Chile. In C. Smith-Ramírez, J. J. Armesto, & C. Valdovinos (Eds.), *Historias, biodiversidad y ecologia de los bosques costeros de Chile* (pp. 505–515). Editorial Universitaria.
Medina-Vogel, G. (2010). Ecology of emerging infectious diseases and wild species conservation. *Archivos de Medicina Veterinaria, 42*, 11–24. https://doi.org/10.4067/s0301-732x2010000100003
Medina-Vogel, G., Delgado Rodríguez, C., Alvarez, P. R. E., & Bartheld, V. J. L. (2004). Feeding ecology of the marine otter (Lutra felina) in a rocky seashore of the south of Chile. *Marine Mammal Science, 20*, 134–144. https://doi.org/10.1111/j.1748-7692.2004.tb01144.x
Medina-Vogel, G., Bartheld, J. L., Pacheco, R. A., & Rodríguez, C. D. (2006). Population assessment and habitat use by marine otter Lontra felina in southern Chile. *Wildlife Biology, 12*, 191–199. https://doi.org/10.2981/0909-6396(2006)12[191:PAAHUB]2.0.CO;2
Medina-Vogel, G., Boher, F., Flores, G., et al. (2007). Spacing behavior of marine otters (*Lontra felina*) in relation to land refuges and fishery waste in Central Chile. *Journal of Mammalogy, 88*, 487–494. https://doi.org/10.1644/06-MAMM-A081R1.1
Medina-Vogel, G., Merino, L. O., & de Monsalve Alarcón, R. J. (2008). Coastal-marine discontinuities, critical patch size and isolation: Implications for marine otter conservation. *Animal Conservation, 11*, 57–64. https://doi.org/10.1111/j.1469-1795.2007.00151.x
Medina-Vogel, G., Calvo-Mac, C., Delgado-Parada, N., Molina-Maldonado, G., Johnson-Padilla, S., & Berland-Arias, P. (2023). Co-occurrence between salmon farming, alien American mink (*Neogale vison*), and endangered otters in Patagonia. *Aquatic Mammals, 49*(6), 561–568. https://doi.org/10.1578/AM.49.6.2023.561
Molina, J. I. (1782). *Saggio sulla storia civile del Chili. Del signor abate Giovanni Ignazio Molina. nella stamperia di S.* Tommaso d'Aquino.
Morrison, P., Rosenmann, M., & Estes, J. A. (1974). Metabolism and Thermoregulation in the Sea Otter. *Physiological Zoology*, 218–229. https://doi.org/10.1086/PHYSZOOL.47.4.30152525

Museo Larco. (n.d.-a). *Botella doble pico asa puente cintada escultórica representando nutria (Número de catálogo: ML019655, Código de ubicación: 169-004-008*). Museo Larco. https://www.museolarco.org/catalogo/ficha.php?id=20805
Museo Larco. (n.d.-b). *Botella gollete asa estribo escultórica representando nutria con cara de felino*. (Número de catálogo: ML008253, Código de ubicación: 093-002-001). Lima, Peru: Museo Larco. https://www.museolarco.org/catalogo/ficha.php?id=8253
Olguín, L., Castro, V., Castro, P., Peña-Villalobos, I., Ruz, J., & Santander, B. (2015). Exploitation of faunal resources by marine hunter–gatherer groups during the Middle Holocene at the Copaca 1 site, Atacama Desert coast. *Quaternary International, 373*, 4–16.
Ortiz, R. M. (2001). Osmoregulation in marine mammals. *The Journal of Experimental Biology, 204*, 1831–1844. https://doi.org/10.1242/JEB.204.11.1831
Ortiz-Alvarez, C., Alfaro-Shigueto, J., & Mangel, J. C. (2021). Insights into marine otter (Lontra felina) distribution along the Peru coastline. *Marine Mammal Science, 37*(3), 993–1007.
Osgood, W. H. (1943). *The mammals of Chile*.
Ostfeld, R., Ebensperger, L., Klosterman, L., & Castilla, J. C. (1989). Foraging, activity budget, and social behavior of the south American marine otter, Lutra felina (Molina, 1782). *National Geographic Research, 5*, 422–438.
Parera, A. (2002). *Los mamíferos de la Argentina y la región austral de Sudamerica*. El Ateneo.
Perrin, M. R., & D'inzilio, I. (1999). Capture, immobilization and measurements of the spotted-necked otter in the Natal Drakensberg, South Africa. *South African Journal of Wildlife Research, 29*, 1–2.
Pizarro-Neyra, J. (2008). Mortality of the marine otter (Lontra felina) in southern Peru. *IUCN Otter Specialist Group Bulletin, 25*(2), 94–99.
Pizarro-Neyra, J. (2022). ¿Ansumo, huallaque o chungungo?, precisiones sobre los nombres comunes de la nutria marina en el Peru1. *Boletín de la Academia Peruana de la Lengua, 71*, 465–476. https://doi.org/10.46744/BAPL.202201.015
Poblete, A. A., Gorski, K., & Moscoso, J. (2019). Estimation of the diet of chungungo Lontra felina (Molina, 1782) in two locations of the coast of the biobío region, Chile. *Gayana, 83*, 1–9.
Pohle, H. (1920). *Die Unterfamilie der Lutrinae*.
Redford, K. H., & Eisenberg, J. F. (1992). *Mammals of the Neotropics, the southern cone: Chile, Argentina, Uruguay, Paraguay*. University of Chicago Press.
Rosello, E., Vasquez, V., Morales, A., & Rosales, T. (2001). Marine resources from an urban Moche (470–600 AD) area in the 'Huacas del sol y de la Luna' archaeological complex (Trujillo, Peru). *International Journal of Osteoarchaeology, 11*(1–2), 72–87.
Rozzi, R., & Torres-Mura, J. C. (1990). Observaciones del Chungungo (Lutra felina) al sur de la Isla Grande de Chiloe: antecedentes para su conservación. *Medio Ambient, 11*, 24–28.
Ruiz, E. (2009). *Estudio etológico de la nutria marina Lontra felina (Molina, 1782) en un ambiente antrópico, la Bahía de Pucusana-Lima, Peru de junio a agosto del 2008*. Universidad Nacional Agraria La Molina (UNALM).
Sánchez, R. (1992). *Distribución y abundancia del gato marino Lutra felina en el Departamento de Ica*.
Santillán, L., & Caro, K. (2007). Mamíferos marinos en el Puerto de Salaverry y Zonas Adyacentes. In *Libro de resúmenes*. CONCIMAR.
Santillán, L., Saldaña-Serrano, M., & De-La-Torre, G. E. (2020). First record of microplastics in the endangered marine otter (Lontra felina). *Mastozoología Neotropical, 27*, 211–215. https://doi.org/10.31687/saremMN.20.27.1.0.12
Schaeffer, N. E. (2006). Fishermen, farmers and fiestas: Continuity in ritual of *traditional villages on the Northwest Coast of Peru*.
Schinz, H. R. (1844). *Systematisches Verzeichniß aller bis jetzt bekannten Säugethiere oder Synopsis mammalium nach dem Cuvier'schen System*. Erster Band.
Schweigger, E. (1964). *El litoral peruano, 2da edició*. Universidad Nacional Federico Villareal.
Secretaría de Ambiente y Desarrollo Sustentable de la Nación y Sociedad Argentina para el Estudio de los Mamíferos. (2019). *Categorización 2019 de los mamíferos de Argentina según su riesgo de extinción*. http://cma.sarem.org.ar/. Accessed 20 Apr 2023.

Sepúlveda, M. A., Singer, R. S., Silva-Rodríguez, E. A., et al. (2014). Invasive American mink: Linking pathogen risk between domestic and endangered carnivores. *EcoHealth, 11*, 409–419. https://doi.org/10.1007/s10393-014-0917-z

Serfass, T., Brooks, R., Swimley, T., et al. (1996). Considerations for capturing, handling, and translocating river otters. *Wildlife Society Bulletin, 24*.

Sielfeld, W. (1983). *Mamíferos marinos de Chile*. Ediciones de la Universidad de Chile.

Sielfeld, W. H. (1989). Sobreposicion de nicho y patrones de distribucion de Lutra felina y L. provocax (Mustelidae, Carnivora) en el medio marino de Sudamerica Austral. *Anales del Museo de Historia Natural de Valparaiso, 20*, 103–108.

Sielfeld, W. (1990a). Características del hábitat de Lutra felina (Molina) y L.provocax Thomas (Carnivora, Mustelidae) en Fuego-Patagonia. *Investig en Cienc y Tecnol Ser Ciencias del Mar, 1*, 30–36.

Sielfeld, W. K. (1990b). Dieta del Chungungo (Lutra felina (Molina, 1782) Mustelidae, Carnivora) en Chile austral. *Investig en Cienc y Tecnol Ser Ciencias del Mar, 1*, 23–29.

Sielfeld, W. K. (1992). Abundancias relativas de Lutra felina (Molina, 1782) y L. provocax (Thomas, 1908) en el litoral de Chile austral. *Investig en Cienc y Tecnol Ser Ciencias del Mar, 2*, 3–11.

Sielfeld, W., & Castilla, J. (1999). Estado de conservación y conocimiento de las nutrias en Chile scientific. *Estud Ocean, 18*, 69–79.

Soto-Azat, C., Boher, F., Flores, G., et al. (2006). Reversible anesthesia in wild marine otters (Lontra felina) using ketamine and medetomidine. *Journal of Zoo and Wildlife Medicine, 37*, 535–538. https://doi.org/10.1638/05-067.1

Spelman, L. H., Symmer, P. W., Levine, J. F., & Stoskopf, M. K. (1994). Immobilization of north American River otters (Lutra Canadensis) with Medetomidine-ketamine and reversal by Atipamezole. *Journal of Zoo and Wildlife Medicine, 25*, 214–223.

Steinel, A., Parrish, C. R., Bloom, M. E., & Truyen, U. (2001). Parvovirus infections in wild carnivores. *Journal of Wildlife Diseases, 37*, 594–607. https://doi.org/10.7589/0090-3558-37.3.594

Tello, E. (1972). Anotaciones sobre el Camarón. *Doc Minist Pesq, 18*, 5–9.

Thomas, O. (1908). LXIII.— On certain African and S.-American otters. *Annals and Magazine of Natural History, 1*, 387–395. https://doi.org/10.1080/00222930808692423

Thomas, O. (1920). Mammalia collected in Peru. *Proc US natn Mus, 58*, 225.

Tschudi, J. (1844). *Reiseskizzen aus Peru*.

Ugarte-Núñez, J. A. (2021). La nutria marina (*Lontra felina*) en los Andes del sur del Peru. *Cienc Desarro, 20*, 71–77. https://doi.org/10.33326/26176033.2021.1.1109

Valqui, J. (2004). *Comportamiento de la nutria marina Lontra felina (Molina, 1782) en una ambiente antrópico, la bahía de Pucusana -Lima*. Universidad Nacional Agraria La Molina (UNALM).

Valqui, J. (2012). The marine otter Lontra felina (Molina, 1782): A review of its present status and implications for future conservation. *Mammalian Biology, 77*, 75–83. https://doi.org/10.1016/j.mambio.2011.08.004

Valqui, J., Hartl, G. B., & Zachos, F. E. (2010). Non-invasive genetic analysis reveals high levels of mtDNA variability in the endangered south-American marine otter (Lontra felina). *Conservation Genetics, 11*, 2067–2072. https://doi.org/10.1007/S10592-010-0069-9/FIGURES/2

Vásquez, V., & Rosales, T. (2016). Análisis de restos de fauna y botánicos de Ca45, Ca52, Ca53, Ca55, Ca56 Y Ca57-Huaca de la Luna, Temporada 2015. In S. Uceda, R. Morales, & C. Rengifo (Eds.), *Investigaciones en la Huaca de la Luna 2015* (pp. 261–287). Facultad de Ciencias Sociales de la Universidad Nacional de Trujillo/Patronato Huacas del Valle de Moche.

Vaz-Ferreira, R. (1979). Situación poblacional y conservación de los mamíferos marinos en Latino America. *Acta Zoológica Lilloana, 34*, 91–101. 10.2/JQUERY.MIN.JS.

Vega-Feria, K., Calvo-Mac, C., Ortiz-Alvarez, C., & Serrano-Martínez, E. (2018). Endoparásitos en heces de nutria marina (Lontra felina) en la costa de Peru. In *XII Congreso de la Sociedad Latinoamericana de Especialistas en Mamíferos Acuáticos – RT* (Vol. 18, p. 335).

Viacava Campos, M., Aitken, S. R., & Llanos Urbina, J. (1978). Estudio del Camaron en el Peru 1975–1976. *Boletin Instituto del Mar del Perú, 3*, 161–232.

Vianna, J. A., Ayerdi, P., Medina-Vogel, G., et al. (2010). Phylogeography of the marine otter (Lontra felina): Historical and contemporary factors determining its distribution. *The Journal of Heredity, 101*, 676–689. https://doi.org/10.1093/jhered/esq088

Villegas, M. J., Aron, A., & Ebensperger, L. A. (2007). The influence of wave exposure on the foraging activity of marine otter, Lontra felina (Molina, 1782) (Carnivora: Mustelidae) in northern Chile. *Journal of Ethology, 25*, 281–286. https://doi.org/10.1007/s10164-006-0024-x

Waterhouse, G. R. (1839). Mammalia. In C. Darwin (Ed.), *La zoología del viaje del HMS Beagle, durante los años 1832 a 1836* (p. 97). Smith Elder and Co.

Williams, T. M., & Worthy, G. A. J. (2002). Anatomy and physiology: The challenge of aquatic living. In A. R. Hoelzel (Ed.), *Marine mammal biology. An evolutionary approach* (pp. 73–97). Blackwell Science.

Chapter 3
Trophic Ecology of the Marine Otter (*Lontra felina*): Review, Quantitative Analyses and Population Implications

Leonardo Hostos-Olivera and Juan Valqui

Abstract The marine otter, *Lontra felina*, is one of the top predators of the benthic communities in the southeastern Pacific coasts, from Chimbote in Peru to Cape Horn in Chile. Despite its importance in these coastal food webs, few studies exist to date on its trophic ecology and an up-to-date comprehensive understanding is needed. A systematic review was thus done on the trophic ecology of *L. felina*, gathering data to address diet composition, feeding patterns and the role of diet on population variability throughout its distribution. The diet of the marine otter comprises a broad spectrum of prey items, including fish, crustaceans, mollusks, birds, echinoderms, mammals, insects, tunicates and even fruits, supporting its classification as a dietary generalist and opportunist. Quantitative analyses on dietary variation across its range revealed a feeding-related biogeographical pattern, consisting of a latitudinal gradient with a higher consumption of fish to the north and crustaceans to the south. Moreover, the feeding ecomorphology of *L. felina* shows that this variation in diet is linked to the phenotypic characteristics of the populations. Northern marine otters, with a higher consumption of fish, exhibit a piscivory-associated morphology (narrower skulls and longer rostrums), while southern marine otters, with a higher consumption of crustaceans, present a durophagy-associated morphology (wider skulls and shorter rostrums). Considering the genetic differences between both populations and the trophic and morphological variations described, it is possible to suggest the existence of a divergence scenario between northern and southern populations of the marine otter, whose knowledge is highly relevant for its conservation given its current status as an endangered species. Despite this, the lack of information on different topics is highlighted, which traces the long road ahead in the study of the trophic ecology of *L. felina*.

L. Hostos-Olivera (✉) · J. Valqui
Facultad de Ciencias y Filosofía/Centro de Investigación para el Desarrollo Integral y Sostenible, Laboratorios de Investigación y Desarrollo, Universidad Peruana Cayetano Heredia, Lima, Peru
e-mail: luis.hostos.o@upch.pe; juan.valqui@gab-kiel.org

L. Ayala et al. (eds.), *Marine Otter Conservation*,
https://doi.org/10.1007/978-3-031-53931-2_3

Keywords Marine otter · Trophic ecology · Diet · Feeding patterns · Ecomorphology

3.1 Introduction

Trophic ecology of an animal species encompasses a unique set of inter and intra-specific characteristics that largely defines its ecological niche at different levels of organization (Garvey & Whiles, 2017). Among the main trophic properties, diet stands out as a critical and very useful source of information. For example, knowledge about the diet of a species enables to compare it with the availability of its prey (De La Vega et al., 2016), assess competition and predator-prey relationships (Spitz et al., 2014; Newsome et al., 2015), evaluate spatial and temporal variation in diet (Díaz-Ruiz et al., 2013; Yurkowski et al., 2016) or even quantify the relative contribution of prey in terms of energetics (Slauson & Zielinski, 2017). Moreover, within the ecological niche context, diets allow to determine the feeding or dietary niche of consumers, which is an integral component that includes the diversity of prey items eaten by a species (Roughgarden, 1972). Thus, diet data represent a fundamental aspect to understand factors that limit populations and, therefore, could serve as a key point for designing conservation management strategies (Kruuk, 2006). Fortunately, the diet and feeding of otters is the aspect of their ecology that has been most thoroughly studied (Kruuk, 2006), although the amount of available information depends on the species in question.

The marine otter, *Lontra felina*, is a particularly interesting otter species from an ecological and evolutionary point of view, as it appears to be the most recently adapted mammal to the marine environment (Estes, 1989). Its habitat is distributed along the Pacific coasts in South America, from Chimbote in Peru to Cape Horn in Chile, and is composed by rocky shore patches, sandy beaches, islands and the nearest 150 meters of the cold and productive waters of the Humboldt Current System (HCS) (Medina-Vogel et al., 2007). Despite its presumed complete adaptation to the marine ecosystem, *L. felina* has also been reported inhabiting freshwater bodies in the Andes of southern Peru (Ugarte-Núñez, 2021), probably reflecting its recent divergence from the river otter *L. provocax* (Vianna et al., 2010).

Regarding trophic relationships, the marine otter is considered one of the top predators of the benthic communities in the marine environments of the HCS, covering both the intertidal and upper subtidal zones (Castilla & Bahamondes, 1979; Soto, 1996; Medina-Vogel et al., 2004; Córdova et al., 2009; Biffi & Iannacone, 2010; Iriarte & Jaksic, 2012; Giacaman-Smith et al., 2016). For that reason, it should play an important and clue role within these coastal food webs, similar to how other otter species do in the aquatic ecosystems in which they inhabit (Kruuk, 2006). However, few studies exist to date on its trophic ecology and the available information, including recent research addressing new contexts, has not been compiled or summarized.

In this chapter, a systematic review of scientific studies on the trophic ecology of *L. felina* was carried out, in order to answer three main questions: (1) What preys compose the diet of the marine otter? (2) How does its diet vary throughout its distribution? (3) Is its feeding spectrum linked to the phenotypic characteristics of populations? These data integration allows us to have a clearer and updated framework about the trophic ecology of the marine otter, understanding its diet composition, the presence of feeding-related biogeographical patterns and the role of diet in intraspecific variability.

3.2 Literature Compilation

We conducted a keyword search on the Google Scholar database to locate published and unpublished studies in English and Spanish that contained information about the trophic ecology of the marine otter, using the following key words: "*Lontra felina*", "marine otter", "diet", "trophic ecology", "chungungo", "nutria marina", "dieta" and "ecología trófica".

In total, we compiled 24 original studies, spanning the period 1979–2022, among which we had 14 research articles, three bachelor thesis, three Congress presentations, two research notes, one brief report and one technical report (see Table 3.1). However, there is an unbalanced result distribution since most of the studies were carried out solely in Chile (17 studies), a few covered areas of the central and southern coast of Peru (5 studies) and only two were conducted considering localities from both geographic regions. The trophic ecology topics that were treated in the compiled studies were diet, feeding behavior, trophic relationships and interactions, and feeding ecomorphology. Finally, in all these studies, four methods were used to collect feeding data: direct observation (12 studies, 48%), fecal analysis (14 studies, 56%), food remains (5 studies, 20%) and stomach content analysis (one study, 4%). Besides that, geometric morphometric techniques were applied for morphological data collection in the ecomorphological study and literature review served to gather information in some other studies and perform subsequent analyses.

3.3 Diet Composition

Based on 15 studies from those shown above, we were able to build a compiled list—and updated until November 2022—of the prey items that are known to compose the diet of *Lontra felina* (see Table 3.2). It should be noted that some of the listed prey items were added based on personal information shared in Juan Valqui's review article on the current status of the marine otter (Valqui, 2012). Furthermore, although all of these studies present prey identifications at the species level, some items were found to be incompletely identified in 12 of them, being described at higher taxonomic levels.

Table 3.1 Review of trophic ecology studies in the marine otter *Lontra felina*

Reference	Region	Topics	Methods	Type
Castilla and Bahamondes (1979)	Chile	Diet, behavior	DO, FA, FR	Research article
Ostfeld et al. (1989)	Chile	Diet, behavior	DO, FA	Research article
Rozzi and Torres-Mura (1990)	Chile	Diet, behavior	FA, FR, SCA	Research article
Sielfeld (1990)	Chile	Diet	FA	Research article
Bernal (1992)	Chile	Diet	DO, FA	Technical report
Soto (1996)	Chile	Trophic relationships	LR	Research article
Ebensperger and Botto-Mahan (1997)	Chile	Trophic relationships	LR	Research article
Villegas et al. (2001)	Chile	Diet	FA	Congress presentation
Mattern et al. (2002)	Chile	Interactions	DO	Brief report
Medina-Vogel et al. (2004)	Chile	Diet	DO, FA, FR	Research article
Valqui (2004)	Peru	Diet, behavior	DO	Thesis
Villegas et al. (2007)	Chile	Behavior	DO	Research article
Biffi and Iannacone (2010)	Peru	Diet	FA	Research article
Mangel et al. (2011)	Peru and Chile	Diet and interactions	DO, FA	Research article
Nuñez-Placencia (2014)	Chile	Diet	FA	Thesis
Pizarro-Neyra (2014)	Peru	Diet, behavior	DO	Congress presentation
Córdova and Rau (2016)	Chile	Diet	FA, FR	Research article
Giacaman-Smith et al. (2016)	Chile	Trophic relationships	LR	Research article
Sanino and Meza (2016)	Chile	Diet, trophic relationships	DO, FA, FR	Research article
Ruelas-Cabana et al. (2018)	Peru	Diet	FA	Congress presentation
Poblete et al. (2019)	Chile	Diet	FA	Research article
Hostos-Olivera (2021)	Peru and Chile	Ecomorphology	GM	Thesis
Rau et al. (2021)	Chile	Diet	DO	Research note
Pizarro-Neyra (2022)	Peru	Interactions	DO	Research note

DO direct observation, *FA* fecal analysis, *FR* food remains, *SCA* stomach content analysis, *LR* literature review, *GM* geometric morphometrics

All the prey item records on the list were made using different data collection methods. Since each method has advantages and disadvantages for prey identification, the combined use of them allows to get a greater scope to know the diet composition of the species. For instance, direct observation of dietary behavior does not always make it possible to completely identify all prey due to observation distance or ingestion speed, however, it can help to record trophic interactions with prey not

Table 3.2 List of prey items in the diet of *Lontra felina* based on literature

Taxonomic groups		Species
FISH		
Order Acanthuriformes	Family Sciaenidae	*Cilus gilberti*
		Sciaena deliciosa
Order Atheriniformes	Family Atherinidae	*Basilichthys semotilus*
	Family Atherinopsidae	*Odontesthes regia*
Order Batrachoidiformes	Family Batrachoididae	*Aphos porosus*
Order Blenniiformes	Family Blenniidae	*Scartichthys viridis*
	Family Clinidae	*Myxodes viridis*
	Family Labrisomidae	*Auchenionchus microcirrhis*
		Auchenionchus variolosus
		Calliclinus geniguttatus
		Labrisomus philippii
Order Carangiformes	Family Carangidae	*Trachurus murphyi*
Order Centrarchiformes	Family Aplodactylidae	*Aplodactylus punctatus*
	Family Cheilodactylidae	*Cheilodactylus variegatus*
	Family Kyphosidae	*Girella laevifrons*
		Graus nigra
Order Clupeiformes	Family Clupeidae	NI
Order Gadiformes	Family Merlucciidae	*Merluccius gayi*
Order Gobiesociformes	Family Gobiesocidae	*Sicyases sanguineus*
Order Ophidiiformes	Family Ophidiidae	*Genypterus chilensis*
		Genypterus maculatus
Order Perciformes	Family Blenniidae	*Hypsoblennius* sp.
	Family Bovichtidae	*Bovichtus chilensis*
		Cottoperca gobio
	Family Eleginopsidae	*Eleginops maclovinus*
	Family Haemulidae	*Isacia conceptionis*
	Family Harpagiferidae	*Harpagifer bispinis*
	Family Nototheniidae	*Patagonothen* sp.
	Family Percidae	NI
	Family Pinguipedidae	*Pinguipes chilensis*
		Prolatilus jugularis
	Family Pomacentridae	*Chromis crusma*
		Nexilosus latifrons
	Family Sebastidae	*Sebastes capensis*
		Sebastes oculatus
	Family Serranidae	*Acanthistius pictus*
	Family Stromateidae	*Stromateus stellatus*
Order Pleuronectiformes	Family Paralichthyidae	*Paralichthys adspersus*
		Paralichthys microps
Order Scombriformes	Family Scombridae	*Katsuwonus pelamis*
		Scomber japonicus

(continued)

Table 3.2 (continued)

Taxonomic groups		Species
Order Siluriformes	Family Ariidae	*Galeichthys peruvianus*
CRUSTACEANS		
Order Amphipoda	Family Gammaridae	*Gammarus* sp.
Order Balanomorpha	Family Balanidae	*Balanus laevis*
Order Decapoda	Family Alpheidae	*Betaeus truncatus*
	Family Belliidae	*Acanthocyclus albatrossis*
		Acanthocyclus gayi
	Family Campylonotidae	*Campylonotus vagans*
	Family Cancridae	*Cancer plebejus*
		Metacarcinus edwardsii
	Family Epialtidae	*Pisoides edwardsii*
		Taliepus dentatus
	Family Grapsidae	*Grapsus*
		Leptograpsus variegatus
	Family Hippidae	*Emerita analoga*
	Family Lithodidae	*Lithodes santolla*
	Family Munididae	*Grimothea gregaria*
		Grimothea monodon
	Family Ovalipidae	*Ovalipes punctatus*
	Family Palaemonidae	*Cryphiops caementarius*
	Family Pilumnoididae	*Pilumnoides perlatus*
	Family Platyxanthidae	*Homalaspis plana*
		Platyxanthus orbignyi
	Family Porcellanidae	*Allopetrolisthes angulosus*
		Allopetrolisthes perlatus
		Allopetrolisthes punctatus
		Allopetrolisthes spinifrons
		Pachycheles crinimanus
		Pachycheles grossimanus
		Petrolisthes desmarestii
		Petrolisthes granulosus
		Petrolisthes tuberculosus
		Petrolisthes violaceus
	Family Rhynchocinetidae	*Rhynchocinetes typus*
	Family Varunidae	*Cyclograpsus cinereus*
		Pseudograpsus setosus
	Family Xanthidae	*Paraxanthus barbiger*
Order isopoda	NI	NI
MOLLUSKS		
Class Bivalvia	Family Mytilidae	*Semimytilus patagonicus*
	Family Mesodesmatidae	*Mesodesma donacium*
	Family Veneridae	*Ameghinomya antiqua*

(continued)

Table 3.2 (continued)

Taxonomic groups		Species
Class Cephalopoda	Family Enteroctopodidae	*Enteroctopus megalocyathus*
	Family Octopodidae	*Octopus mimus*
		Octopus vulgaris
Class Gastropoda	Family Fissurellidae	*Fissurella crassa*
		Fissurella latimarginata
		Fissurella limbata
		Fissurella maxima
	Family Calyptraeidae	*Crepipatella dilatata*
	Family Lottiidae	*Collisella* sp.
		Scurria parasitica
		Scurria scurra
		Scurria viridula
		Scurria zebrina
	Family Muricidae	*Concholepas*
	Family Tegulidae	*Tegula atra*
	Family Lymnaeidae	*Galba viatrix*
Class Polyplacophora	Family Chitonidae	*Chiton cumingsii*
		Chiton granosus
BIRDS		
Order Anseriformes	Family Anatidae	*Chloephaga hybrida*
		Lophonetta specularioides
		Tachyeres pteneres
Order Charadriiformes	Family Laridae	*Larosterna inca*
		Larus dominicanus
		Larus scoresbii
		Sterna hirundinacea
Order Falconiformes	Family Falconidae	*Phalcoboenus chimango*
Order Pelecaniformes	Family Ardeidae	*Nycticorax nycticorax*
Order Procellariiformes	Family Pelecanoididae	*Pelecanoides garnotii*
Order Suliformes	Family Phalacrocoracidae	NI
MAMMALS		
Order Rodentia	Family Cricetidae	*Phyllotis* sp.
ECHINODERMS		
Class Echinoidea	Family Echinidae	*Loxechinus albus*
TUNICATES		
Class Ascidiacea	Family Pyuridae	*Pyura chilensis*
INSECTS		
Order Coleoptera	NI	NI
Order Odonata	Infraorder Anisoptera	NI
PLANTS		
Order Poales	Family Bromeliaceae	*Fascicularia bicolor*
		Greigia sphacelata

NI not identified, at lower taxonomic levels

found by other methods or even to confirm the presence of certain prey items in the diet (Valqui, 2004; Biffi & Iannacone, 2008). Then, fecal analysis is a non-invasive, quick and simple method to characterize the diet through the identification of prey structures that resist digestion and are found in spraints (e.g., otoliths in fish). Although spraints (otter fecal samples) are relatively abundant and easy to collect, a major limitation of the method is its low capacity to register prey that does not leave solid remains, as occurs with mollusks and echinoderms of which only the soft parts are ingested (Biffi & Iannacone, 2008; Nuñez-Placencia, 2014). On the other hand, this methodological problem can be compensated by the method based on food remains left in feeders, since the marine otter leaves there only the remains that cannot be consumed, such as mollusk shells (Nuñez-Placencia, 2014; Córdova & Rau, 2016). Finally, stomach content analysis can also help to confirm the consumption of some prey items, but the difficulty is that individuals have to be sacrificed and many of the stomachs to be examined could even be empty (Biffi & Iannacone, 2008).

In summary, a total of 116 species were identified as part of the diet of the marine otter, including 42 fish, 36 crustaceans, 21 mollusks, 11 birds, 1 mammal, 1 echinoderm, 1 tunicate, 2 insects and 2 plants. Based on this, fish and crustaceans stand out by far as the most diverse prey groups of the diet of *L. felina* (Fig. 3.1).

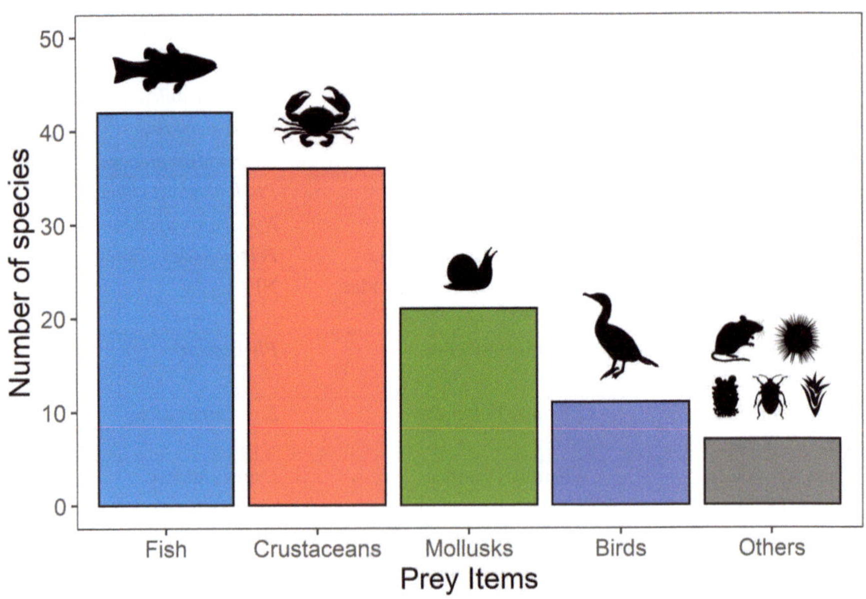

Fig. 3.1 Number of species of prey item categories in the diet of *Lontra felina* based on the literature review. Prey category placed as “Others” includes mammals, echinoderms, tunicates, insects and plants (precisely fruits)

Fish The registered fish species belonged to 30 families and 14 orders, most of them characteristic of the intertidal zone of the marine ecosystem. Three orders concentrated more than 60% of the species: Perciformes (16 species), Blenniiformes (6 species) and Centrarchiformes (4 species). In addition, the Labrisomidae family is distinguished by being one of the most frequent and the most diverse in the diet with four species: *Auchenionchus microcirrhis*, *A. variolosus*, *Calliclinus geniguttatus* and *Labrisomus philippii*. From the list, the fish species most frequently reported as marine otter prey items in diet studies were the Pacific sandperch *Prolatilus jugularis* (locally known as "blanquillo") and the Patagonian redfish *Sebastes oculatus* (locally known as "escrófalo"). Other species to highlight due to their high frequencies of occurrence in the studies were the Patagonian blennie *Eleginops maclovinus*, the Chilean thornfish *Bovichtus chilensis* and the frog-fish or "pejesapo" *Sicyases sanguineus* (Castilla & Bahamondes, 1979; Medina-Vogel et al., 2004; Nuñez-Placencia, 2014). Besides that, the Andean silverside or "pejerrey andino", *Basilichthys semotilus*, is positioned as the only freshwater fish species recorded in the marine otter diet, inhabiting rivers and lagoons at altitudes close to 2000 meters above sea level in southern Peru (Ruelas-Cabana et al., 2018).

Crustaceans From mainly hard parts in feces, it was possible to identify 19 families of crustaceans present in the diet of *L. felina*, belonging to four orders. Of these, the order Decapoda encompassed more than 90% of the species (33 species) and the family of porcelain crabs, Porcellanidae, stood out as the most diverse with 10 species. Most of the recorded crustaceans correspond to the upper rocky littoral, being found from the supratidal zone to between 20 and 30 meters deep in the intertidal and subtidal zones. The crustacean species most frequently reported as prey in the marine otter diet were the hairy crab *Pseudograpsus setosus*, the blackberry crab *Homalaspis plana* and the spider crab *Taliepus dentatus* (locally known as "panchote"). The latter, *T. dentatus*, stands out for usually having the highest frequencies of occurrence both in spraints and food remains, in addition to other species to highlight such as the porcelain crab *Petrolisthes desmarestii* and the mola rock crab *Metacarcinus edwardsii* (Medina-Vogel et al., 2004; Nuñez-Placencia, 2014). Furthermore, crabs are not the only recorded decapods in the diet, there are also lobsters, such as *Grimothea gregaria* and *G. monodon* (Sielfeld, 1990; Nuñez-Placencia, 2014; Poblete et al., 2019), and shrimps, such as *Betaeus truncatus* and *Rhynchocinetes typus* (Ostfeld et al., 1989; Pizarro-Neyra, 2014). Even a distinguishable shrimp species is *Cryphiops caementarius*, which is the most frequent in the diet of the populations of *L. felina* in the Colca-Majes rivers in Peru (Ruelas-Cabana et al., 2018). Besides decapods, three other orders of crustaceans have been registered: Isopoda, Amphipoda and Balanomorpha. Among the records, a species of isopod (not identified at the species level), the freshwater amphipod *Gammarus* sp. and the barnacle *Balanus laevis* are listed.

Mollusks Species of mollusks in the marine otter diet belonged to 12 families, grouped into four classes. The Gastropoda class comprised more than 60% of the species (13 species) and the most diverse families were Lottiidae and Fissurellidae with five and four species, respectively. Most of these species are associated with the rocky substrate in the intertidal zone, some being found at lower benthic levels.

It is worth mentioning that, unlike other prey items that were mainly identified through fecal samples, mollusks were also identified to a large extent by food remains left in the feeders. The mollusk species most frequently reported as marine otter prey items were the Chilean abalone or "loco" *Concholepas concholepas*, the black sea snail *Tegula atra*, the sea cap or "gorrito de mar" *Scurria scurra*, the mussel *Semimytilus patagonicus* and the gould octopus *Octopus mimus*. Among these, *C. concholepas* is remarkable for having one of the highest frequencies of occurrence in the diet of *L. felina*, through food remains, and for being considered a key species in the food web of marine-coastal communities of the HCS in Chile (Giacaman-Smith et al., 2016; Córdova & Rau, 2016). In addition, the pond snail *Galba viatrix* is the only freshwater gastropod recorded in the diet, inhabiting the Mamacocha lagoon at 1700 masl in the Peruvian Andes (Ruelas-Cabana et al., 2018). Besides gastropods and cephalopods, there are other mollusks to highlight that have been found composing the marine otter diet, such as the chitons *Chiton cumingsii* and *Chiton granosus* (class Polyplacophora) and the bivalves *Ameghinomya antiqua* and *Mesodesma donacium* (class Bivalvia).

Birds The predation of bird species by *L. felina* has been considered as secondary and occasional, being recorded mainly through direct observations, spraints and prey rests in dens. Six families of birds, each representing one order, have been recorded as part of its diet. The families Laridae (order Charadriiformes) and Anatidae (order Anseriformes) were the most diverse with four and three species, respectively, having gulls (*Larus dominicanus* and *L. scoresbii*) and terns (*Larosterna inca* and *Sterna hirundinacea*) in the first, and ducks (*Lophonetta specularioides* and *Tachyeres pteneres*) and a goose (*Chloephaga hybrida*) in the second. Other reported bird species were the chimango caracara *Phalcoboenus chimango* (order Falconiformes), the night heron or "huairavo" *Nycticorax nycticorax* (order Pelecaniformes), the Peruvian diving petrel *Pelecanoides garnotii* (order Procellariiformes) and an unidentified species of the family Phalacrocoracidae (order Suliformes) that could probably be the Magellanic cormorant *Leucocarbo magellanicus*. It should be noted that most of the trophic interactions of these species with *L. felina* were reported for the Magallanes region (Sielfeld, 1990), an area where there have been no reports of marine otter populations for decades and is currently considered as potential distribution of the species (Valqui, 2012). Therefore, these records of birds in the diet must be taken with caution, suggestively as historical records. Still, less old records further north such as *P. garnotii* on Choros Island in Chile and *L. inca* on San Gallán Island in Peru (Mattern et al., 2002) validate the presence of birds in the diet, at least as occasional prey. Beyond diet, trophic interactions of the marine otter with birds such as the Peruvian pelican *Pelecanus thagus* have been documented (Pizarro-Neyra, 2022).

Other Prey Among the groups of prey items identified in the diet of *L. felina*, some have a very low diversity (number of species) that would allow them to be considered as circumstantial or even accidental prey. It is necessary to specify that this is not unusual to find in the marine otter, since it presents an opportunistic feeding behavior, meaning that the selection of its prey is based on their availability (Medina-Vogel et al., 2004). On the one hand, the records of mammals such as the rodent *Phyllotis* sp. and insects such as beetles (Coleoptera) and dragonflies (Anisoptera) occurred only in a very particular context, which corresponds to the marine otters in the Mamacocha lagoon (Ruelas-Cabana et al., 2018). This case is a clear example of how *L. felina* adapts its diet in a situation where its main prey items are not as readily available. On the other hand, the red sea urchin *Loxechinus albus* was recorded in a few areas of Chile as part of the marine otter diet through direct observation, food remains, and stomach content analysis (Castilla & Bahamondes, 1979; Rozzi & Torres-Mura, 1990; Rau et al., 2021). However, unlike the North Pacific sea otter *Enhydra lutris*, it appears that the South American marine otter *Lontra felina* does not have urchins among its main prey items as they are not frequently found in diet studies (Kruuk, 2006). Therefore, sea urchins could be considered occasional prey for marine otters, which makes evident the different ecological niches occupied by the two otter species most closely related to the marine habitat (Castilla & Bahamondes, 1979; Bernal, 1992). At last, the singular records of tunicates (such as the "piure" *Pyura chilensis*) and plants (such as the bromeliads *Fascicularia bicolor* and *Greigia sphacelata*) in the diet would indicate a casual or accidental consumption of these prey by individuals of the marine otter, which also date from the last century, so they would be classified as historical records.

3.4 Biogeographical Patterns in Diet

3.4.1 Methods: Data Selection, Spatialization and Analyses

In order to describe the geographic variation in diet composition of *L. felina*, we carried out a standardization of data from different geographical areas that can be used for later comparisons and analysis. With this purpose, we did a selection of diet studies based on three criteria, excluding those: (i) not using fecal analysis as a method for collection of feeding data; (ii) reporting quantitative data for only some of the prey groups found; and (iii) reporting only the frequency of occurrence (FO, expressed as the percentage of spraints containing a particular prey item). These exclusions meant that we only considered studies reporting the relative frequency of occurrence (RFO, expressed as the percentage of times one prey item occurs in relation to the total times all prey items occur) for all the prey groups identified through spraint analysis. The decision to use RFO as the appropriate methodology to establish feeding geographical patterns was based on the fact that, on the one hand, RFO values have been considered to be highly suitable for inter-population comparisons in diet studies (Clavero et al., 2003) and, on the other hand, most of the marine otter studies compiled in this review presented RFO information.

From this selection, a final set of 8 studies resulted, covering 15 localities distributed in southern Peru and throughout the Chilean region (see Table 3.3). Sample

Table 3.3 Information from diet studies of *Lontra felina* based on fecal analysis, throughout its distribution in Peru and Chile

Locality	Region	SS	Duration	Reference
Mamacocha lagoon	Peru	18	2016–2018	Ruelas-Cabana et al. (2018)
Colca-Majes Rivers	Peru	13	2016–2018	Ruelas-Cabana et al. (2018)
Ilo-VilaVila	Peru	101	2003–2009	Mangel et al. (2011)
Pan de Azúcar	Chile	25	1988	Ostfeld et al. (1989)
Punta Cachos	Chile	137	2005–2006	Mangel et al. (2011)
Choros Island	Chile	56	2001	Villegas et al. (2001)
Caleta Arrayán	Chile	136	2005–2006	Mangel et al. (2011)
Totoralillo	Chile	77	2005–2006	Mangel et al. (2011)
San Vicente	Chile	66	2016	Poblete et al. (2019)
Caleta Chome	Chile	65	2016	Poblete et al. (2019)
Valdivia	Chile	475	1999–2000	Medina-Vogel et al. (2004)
Corral	Chile	8	2005–2006	Mangel et al. (2011)
Pucatrihue	Chile	35	2003–2004	Córdova and Rau (2016)
Chiloé Island	Chile	24	1988	Ostfeld et al. (1989)
Guafo Island	Chile	67	2012–2013	Nuñez-Placencia (2014)

SS Sample size (number of spraints collected in each study site)

size and duration of the studies were highly variable, with samples ranging from 8 to 475 spraints and study durations being seasonal, annual or even multi-year, so the results obtained should be interpreted with caution.

For the comparison of the diet composition between localities, we represented the percentages of RFO by locality in the geographic space, distinguishing six prey groups: fish, crustaceans, mollusks, mammals, insects and plants. While this allows us to describe how diet varies along marine otter distribution, it does not make it possible to establish associations between geography and the occurrence of major prey items, with further statistical support. For that reason, we first employed a principal component analysis (PCA) to summarize the variability of RFO values of prey groups in marine-coastal populations of *L. felina* (i.e., avoiding data from Mamacocha lagoon and Colca-Majes rivers) and then we performed regression analyses using latitude (the geographic variable) as a predictor and the first principal components (PC1 and PC2) as response variables. All these analyses were carried out in the R environment (R Core Team, 2022).

3.4.2 *Variation in Diet Composition Across the Distribution*

Through spatialization of RFO values per locality, we found a clear distinction between the diets of the freshwater and marine populations of *L. felina* (Fig. 3.2). The populations found in water bodies of southern Peruvian Andes presented a diet with a predominance of crustaceans (river shrimps) in the Colca-Majes rivers and insects (beetles and dragonflies) and mollusks (snails) in the Mamacocha lagoon.

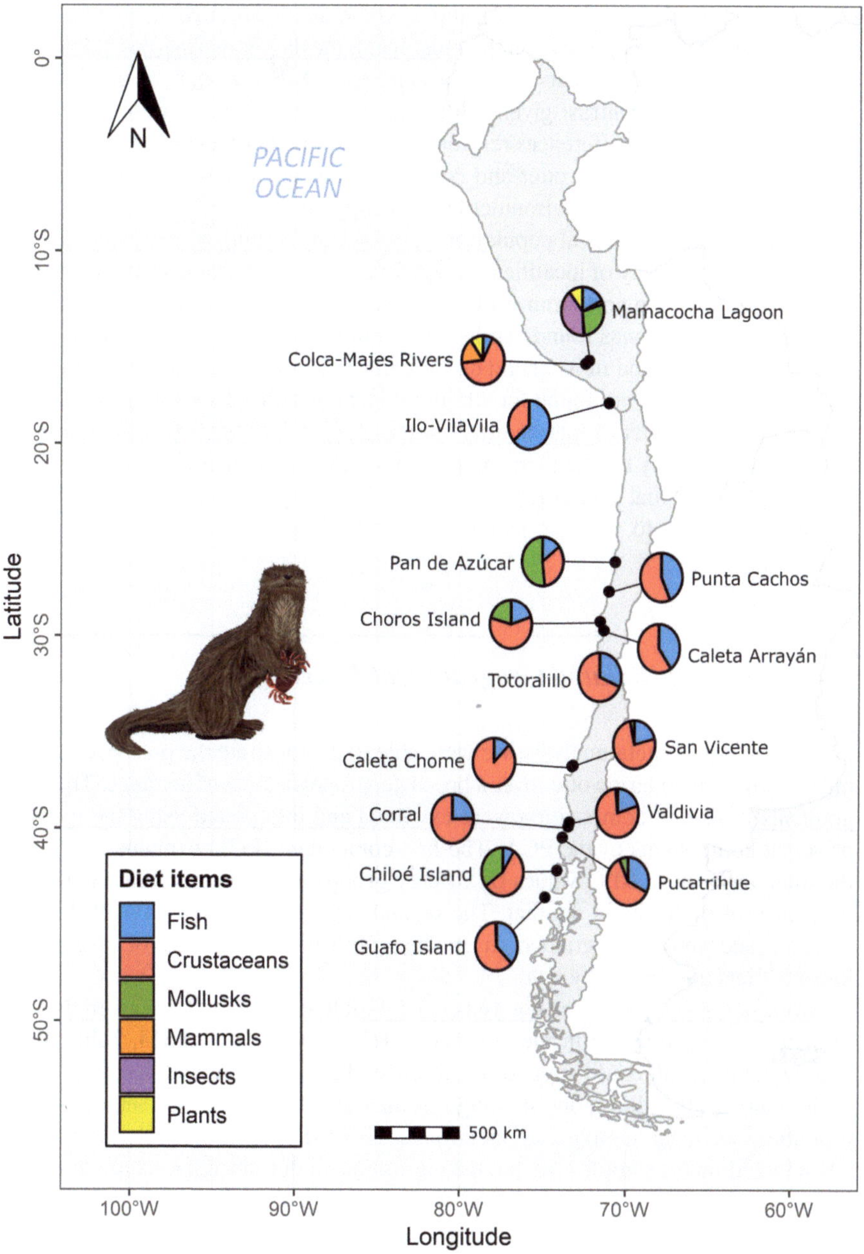

Fig. 3.2 Diet composition of the marine otter *Lontra felina* across localities in Peru and Chile, based on relative frequency of prey occurrence data reported in fecal analysis studies. Information from localities and diet studies are given in Table 3.3

Furthermore, these were the only localities where the occurrence of mammals and plants in the diet was reported. On the other hand, the diet of marine-coastal populations consisted almost exclusively of crustaceans (mostly crabs), fish and mollusks, showing an evident contrast given by the main groups of prey consumed in each type of habitat. These differences reinforce the assignment of an opportunistic feeding behavior to the marine otter and confirm its dietary adaptability based on the available resources in the environment it is occupying.

Within the marine-coastal populations, crustaceans comprised more than 50% of the diet for the majority of localities, except for the two northernmost sites (Fig. 3.2). In Pan de Azúcar, the northernmost Chilean locality with data, a higher percentage of mollusks (51.7%) was found, while to the south of the Peruvian coast from Ilo to Vila Vila, fish were the most abundant prey in the diet (63.9%). Although mollusk occurrence data were not found for all littoral locations, this does not mean that they are not part of the diet of marine otters in these areas, but rather that their absence in the samplings may be due to the major limitation of spraint analysis related to the recording of prey that do not leave solid remains. Despite that, these findings as a whole could point to possible patterns in the feeding of *L. felina* throughout its distribution.

3.4.3 Feeding-Related Geographical Patterns

From quantitative meta-analysis, we were able to demonstrate the presence of feeding patterns in the marine otter over a broad geographical area of its range. The main gradients in its diet composition were extracted and interpreted using the two first principal components of the PCA. The first component (PC1) explained 52.9% of the total variance in RFO values of all prey groups and represented an increasing frequency of mollusks in the diet. The second (PC2) explained 47.1% of the total variance and showed a gradient from diets with high frequency of crustaceans towards diets dominated by fish.

Mollusk consumption was not associated with latitude (Fig. 3.3a) given the lack of statistical significance in the correlation ($F_{1,11} = 0.27$, $p = 0.614$, Table 3.4a). Although this result could suggest the absence of a mollusk-related feeding pattern in the marine otter diet along its range, it may also be due to the methodological limitations of fecal samples, as mentioned above. For instance, this is observed when including the sample size into the regression model, since it not only improves the model fit (R^2 changes from 0.024 to 0.47), but also a significant effect of the interaction between latitude and sample size is found ($F_{1,11} = 6.01$, $p = 0.037$, Table 3.4b), reflecting that the association between latitude and mollusk consumption depends on the number of spraints per locality. Thus, further analyses including occurrence data from food remains should be done to clarify this issue.

Interestingly, a geographic pattern was found regarding the consumption of fish and crustaceans by *L. felina* (Fig. 3.3b). The statistical analysis showed a significant correlation between the latitude and the gradient of fish-crustacean occurrence in

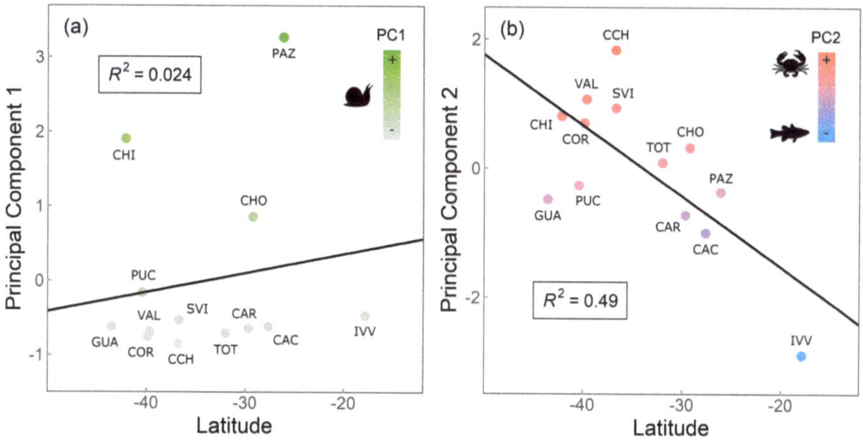

Fig. 3.3 Relationships between latitude and **(a)** PC1 reflecting the variation in mollusk consumption and **(b)** PC2 reflecting a gradient of fish and crustacean consumption by the marine-coastal populations of *Lontra felina*. Each point represents one locality: Ilo-VilaVila (IVV), Pan de Azúcar (PAZ), Punta Cachos (CAC), Choros Island (CHO), Caleta Arrayán (CAR), Totoralillo (TOT), San Vicente (SVI), Caleta Chome (CCH), Valdivia (VAL), Corral (COR), Pucatrihue (PUC), Chiloé Island (CHI), Guafo Island (GUA)

the diet ($F_{1,11} = 10.45$, $p = 0.008$, Table 3.4c), which is not affected by including the sample size in the regression model (Table 3.4d). Marine otters were thus observed to more frequently consume crustaceans in southern regions and more fish in northern regions, as previous studies reported considering fewer locations (Mangel et al., 2011).

Based on diversity and frequency in the diet, crustaceans and fish were clearly the primary prey types for the marine otter, so it makes sense to find that their consumption maintains close relationships with the geographic range of the species. With respect to mollusks, they seem to be frequent but secondary prey in the diet, therefore, it would be less likely to find geographic-food patterns related to them. Nonetheless, it is important to emphasize that most of the localities analyzed belonged to Chile, having data only from one location in southern Peru. Given that, it is necessary to obtain and gather more dietary data on several regions along the species distribution (e.g., north and south of Chiloé Island in Chile, and most of the Peruvian locations) in order to continue supporting the existence of this biogeographical pattern in the diet.

On the other hand, it is interesting to find that the marine otter, *L. felina*, feeds more frequently on crustaceans in the southern regions, being these areas also inhabited by the river otter or "huillín" *Lontra provocax* (Sepúlveda et al., 2021). In fact, it has been reported that the diet of *L. provocax* in the marine environment is mainly composed of coastal fish (Sielfeld, 1992; Ebensperger & Botto-Mahan, 1997). However, the sympatric presence of both species has not yet been demonstrated, since to date there is no clear evidence—only anecdotal (Sanino & Meza, 2016)—of interactions that show either coexistence or exclusion between them.

Table 3.4 Regression analyses results associating latitude, sample size and their interaction with the principal components of diet composition variation of *Lontra felina*. R-squared values are set next to each regression model. Significant *p*-values (< 0.05) are shown in bold

Regression model	DF	SS	MS	*F*	*P*-value
(a) PC1 ~ latitude, $R^2 = 0.024$					
Latitude	1	0.454	0.454	0.27	0.614
Residuals	11	18.576	1.689		
(b) PC1 ~ latitude * sample size, $R^2 = 0.47$					
Latitude	1	0.454	0.454	0.41	0.540
Sample size	1	1.770	1.770	1.58	0.240
Latitude × sample size	1	6.730	6.730	6.01	**0.037**
Residuals	9	10.076	1.120		
(c) PC2 ~ latitude, $R^2 = 0.49$					
Latitude	1	8.268	8.268	10.45	**0.008**
Residuals	11	8.702	0.791		
(d) PC2 ~ latitude * sample size, $R^2 = 0.61$					
Latitude	1	8.268	8.268	11.25	**0.008**
Sample size	1	0.064	0.064	0.09	0.774
Latitude × sample size	1	2.024	2.024	2.75	0.131
Residuals	9	6.615	0.735		

DF degrees of freedom, *SS* sum of squares, *MS* mean squares, *F* value of F statistic

3.5 Feeding Ecomorphology

Variability in feeding habits across populations of a species can be a cause or consequence of phenotypic changes. Given the high dietary plasticity that otter species can exhibit, this is expected to be reflected in modifications of the cranial and mandibular morphology, mainly associated with the dentition and the masticatory muscle attachment area (Kitchener et al., 2017; Meloro & Tamagnini, 2022). In this sense, some intraspecific studies on otter morphology have shown that geographical variation in skull shape is related to diet, so the difference in dietary regimens may favor (Russo et al., 2022) or be favored (Campbell & Santana, 2017) by morphological divergence.

Through studying the feeding ecomorphology of the marine otter *Lontra felina*, Hostos-Olivera (2021) was able to determine a diet-related morphological variation between populations of different geographic regions (Fig. 3.4a). Geometric morphometric analysis of the dorsal (Fig. 3.4b) and ventral (Fig. 3.4c) views of the cranium indicated that contrary morphological traits between northern (Peru, $n = 30$) and southern (Chile, $n = 16$) populations of the marine otter were aligned with the two feeding ecomorphotypes established for otters (Hostos-Olivera, 2021). On the one hand, northern marine otters exhibited proportionally narrower skulls and more distally extended rostrums, typical features of the piscivore ecomorphotype that functionally favor their hydrodynamic shape and give them a greater advantage to capture fast and elusive fish with their canines (Timm-Davis et al., 2015;

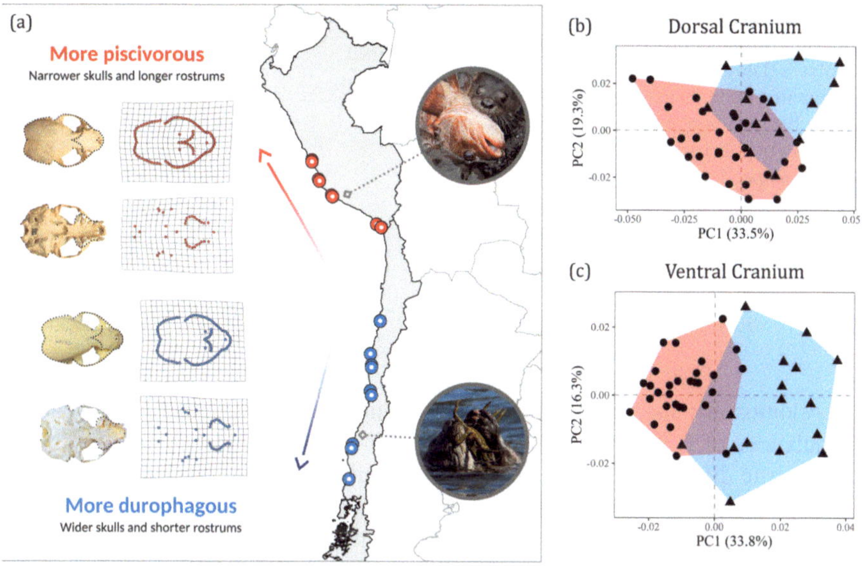

Fig. 3.4 Cranial morphological variation related to differences in diet between northern (red) and southern (blue) populations of *Lontra feline*. **(a)** Map modified from Hostos-Olivera (2021) showing the study sites and the characteristic traits of both populations that align with piscivorous and durophagous ecomorphotypes in otters. Principal component analysis (PCA) of cranial morphology in **(b)** dorsal and **(c)** ventral views among geographic regions

Hostos-Olivera, 2021). On the other hand, southern marine otters exhibited wider skulls and shorter rostrums, characteristic traits of the durophagous ecomorphotype that provide them with greater cranial resistance and bite force to feed more efficiently on hard-bodied invertebrates such as crustaceans (Timm-Davis et al., 2015; Hostos-Olivera, 2021).

Precisely, this significant differentiation in cranial morphology (see Table 3.5) coincides with the geographic pattern described above in relation to the consumption of fish and crustaceans by *L. felina*. Integrating all the variability recorded, we observe that northern populations feed more frequently on fish and shows a piscivorous morphotype, whereas southern populations consume mostly crustaceans and presents a durophagous morphotype. Therefore, it can be established that the variations in geography, morphology and feeding are closely related at the population level in the marine otter, although it is difficult to define an order for these patterns in an evolutionary context and propose causal relationships between them. Still, we can suggest that some scenarios are more likely than others. For example, given that anthropogenic activities have modified the structure of intertidal marine communities (Kunze et al., 2021; Rivadeneira & Nielsen, 2022), probably influencing the diet of *L. felina*, it is more likely that the appearance of the observed population morphological pattern—subject to a longer time scale evolutionary process—has preceded trophic differentiation.

Table 3.5 Results of discriminant function analyses (DFA) for differentiation in cranial shape between northern and southern populations of *Lontra felina*. Mean Procrustes distances and *p*-values, obtained from permutation tests with 10,000 iterations, are shown

Cranial view	Procrustes distance	*P*-value
Dorsal	0.0271	< 0.001
Ventral	0.0204	< 0.001

Additionally, the ecomorphological variation between northern and southern populations of *L. felina* becomes more relevant when considering in the equation the genetic differences found among both population groups. Based on phylogeographic analyses, two divergent haplogroups were revealed, one composed of haplotypes exclusively from Peru and the other with all haplotypes from Chile, which were defined as two evolutionary significant units after identifying the presence of long sandy beaches that act as barriers to dispersal (Vianna et al., 2010). The occupancy of the marine otter depends greatly on the rocky shore patches and the extension of sandy beaches can affect the availability of optimal habitat conditions, such as wave exposure and proximity to food resources (Medina-Vogel et al., 2007; Villegas et al., 2007). Thus, large separation distances due to the extensive beaches and the absence of suitable habitat in them can constitute these areas as permanent physical barriers for the species. As a whole, this information suggests the current existence of a pattern of population divergence in *L. felina* that includes geographical, genetic, ecological (diet) and morphological components. Nevertheless, as for the trophic evaluation, it has not been possible to cover a complete sampling of the marine otter range in the studies of genetic and morphological variability. Thus, it is essential to increase the specimen sampling, especially for areas without data, to confirm or reject the patterns already described.

3.6 Conclusions and Future Directions

The systematic review on the trophic ecology of the South American marine otter, *Lontra felina*, allowed us to confirm that its diet comprises a broad spectrum of food items, including fish and crustaceans as the primary prey, and mollusks, birds, echinoderms, mammals, insects, tunicates, and even fruits as secondary or circumstantial prey. Based on this, its classification as a dietary generalist and opportunist is supported. However, in order to consolidate this statement, it will be necessary to collect new ecological data and compare the proportions of consumed prey with their observed abundances in environments occupied by marine otters. In fact, there is only one study to date that attempted to evaluate diet compared to prey availability (Medina-Vogel et al., 2004), which demonstrates that the trophic ecology of *L. felina* is still very unknown.

Quantitative analyses on the variation in diet composition across its range reflected a latitudinal gradient in the consumption of fish and crustaceans. A higher dietary occurrence of fish to the north and crustaceans to the south was found, thus revealing an evident feeding-related biogeographical pattern. Even so, analyses that incorporate more dietary data from Peruvian and southern Chilean populations are needed, therefore, it is suggested to promote the collection of spraint samples of marine otters to the north and south of the distribution. Moreover, the incorporation of novel technologies such as stable isotope analysis and DNA metabarcoding would improve dietary characterizations (Whitaker et al., 2019), achieving a better understanding of the trophic ecology of the marine otter.

The study of the feeding ecomorphology of *L. felina* enables to link its dietary variation with intrinsic characteristics of the populations such as phenotypic traits. Northern marine otters, with a higher consumption of fish, exhibits a piscivory-associated morphology (narrower skulls and longer rostrums), while southern marine otters, with a higher consumption of crustaceans, presents a durophagy-associated morphology (wider skulls and shorter rostrums). Considering the genetic differences between both populations groups (Vianna et al., 2010), these findings shed light on a multi-dimensional pattern of population divergence in *L. felina*, which has evolutionary implications and whose knowledge is of relevance for its conservation due to its current status as an endangered species (Mangel et al., 2022). However, in the absence of representative sampling for the entire range of the species, we limit ourselves to suggest the existence of this possible scenario of intraspecific variability and to evidence the scarcity of information, pointing out the long road ahead in the study of the trophic ecology of *L. felina*.

Besides all the above, among the trophic relationships of *L. felina* that need special attention for research is the one associated with parasites. Given that a large number of its prey (mainly fish) function as intermediate hosts of endoparasites for various carnivore species, marine otters have a potential pathogenicity as definitive hosts (Molina-Maldonado et al., 2022), which certainly affects their survival. For this reason, it is recommended to prioritize the identification of the endoparasite species found in the marine otter and the evaluation of their prevalence (Calvo-Mac et al., 2020; Molina-Maldonado et al., 2022), contributing to the study of one of the most relevant trophic ecology issues currently for its conservation.

References

Bernal, R. (1992). Conducta, hábitat y dieta alimentaria del chungungo, *Lutra felina*, Molina, 1782 (Carnivora, Mustelidae) en el islote de Cachagua. *Antecedentes de los Principales Recursos Naturales del Monumento Natural Isla Cachagua, V Región*, 14–25.

Biffi, D., & Iannacone, J. (2008). Conservación y ecología trófica de *Lontra felina* (Molina, 1782) (Carnivora: Mustelidae) en el Perú. *Biologist, 6*, 72–76.

Biffi, D., & Iannacone, J. (2010). Variabilidad trófica de *Lontra felina* (Molina, 1782) (Carnivora: Mustelidae) en dos poblaciones de Tacna (Perú) entre agosto y diciembre de 2006. *Mastozoología Neotropical, 17*, 11–17.

Calvo-Mac, C., Gutleb, A. C., Contal, S., et al. (2020). Exposure to toxoplasma gondii in marine otters (*Lontra felina*) and domestic cats (*Felis catus*) in an arid environment in Chile. *Journal of Wildlife Diseases, 56*, 962–964.

Campbell, K. M., & Santana, S. E. (2017). Do differences in skull morphology and bite performance explain dietary specialization in sea otters? *Journal of Mammalogy, 98*, 1408–1416.

Castilla, J. C., & Bahamondes, I. (1979). Observaciones conductuales y ecológicas sobre *Lutra felina* (Molina, 1782) (Carnivora: Mustelidae) en las zonas Central y Centro-Norte de Chile. *Archivos de Biología y Medicina Experimentales (Santiago), 12*, 119–132.

Clavero, M., Prenda, J., & Delibes, M. (2003). Trophic diversity of the otter (*Lutra lutra L.*) in temperate and Mediterranean freshwater habitats. *Journal of Biogeography, 30*, 761–769.

Córdova, O., & Rau, J. (2016). Interacción entre la pesca artesanal y el depredador de alto nivel trófico *Lontra felina* en Chile. *Revista de Biología Marina y Oceanografía, 51*, 621–627.

Córdova, O., Rau, J., Suazo, C., & Arriagada, A. (2009). Estudio comparativo de la ecología alimentaria del depredador de alto nivel trófico *Lontra felina* (Molina, 1782) (Carnivora: Mustelidae) en Chile. *Revista de Biología Marina y Oceanografía, 44*, 429–438.

De La Vega, C., Lebreton, B., Siebert, U., et al. (2016). Seasonal variation of harbor Seal's diet from the wadden sea in relation to prey availability. *PLoS One, 11*, 1–21.

Díaz-Ruiz, F., Delibes-Mateos, M., García-Moreno, J. L., et al. (2013). Biogeographical patterns in the diet of an opportunistic predator: The red fox *Vulpes vulpes* in the Iberian Peninsula. *Mammal Review, 43*, 59–70.

Ebensperger, L., & Botto-Mahan, C. (1997). Use of habitat, size of prey, and food-niche relationships of two sympatric otters in southernmost Chile. *Journal of Mammalogy, 78*, 222–227.

Estes, J. A. (1989). Adaptations for aquatic living by carnivores. In J. L. Gittleman (Ed.), *Carnivore behavior, ecology and evolution* (pp. 242–282). Cornell University Press.

Garvey, J. E., & Whiles, M. R. (2017). *Trophic ecology* (1st ed.). CRC Press, Taylor & Francis Group.

Giacaman-Smith, J., Neira, S., & Arancibia, H. (2016). Community structure and trophic interactions in a coastal management and exploitation area for benthic resources in Central Chile. *Ocean and Coastal Management, 119*, 155–163.

Hostos-Olivera, L. (2021). *Ecomorfología de la alimentación de la nutria marina Lontra felina (Molina, 1782) (Carnivora: Mustelidae)*. Tesis de Licenciatura. Departamento de Ciencias Biológicas y Fisiológicas, Universidad Peruana Cayetano Heredia.

Iriarte, A., & Jaksic, F. (2012). *Lontra felina*. In A. Iriarte & F. Jaksic (Eds.), *Los Carnívoros de Chile* (pp. 168–175). Flora & Fauna Chile.

Kitchener, A. C., Meloro, C., & Williams, T. M. (2017). Form and function of the musteloids. In D. W. Macdonald, C. Newman, & L. A. Harrington (Eds.), *Biology and conservation of Musteloids* (pp. 92–128). Oxford University Press.

Kruuk, H. (2006). *Otters: Ecology, behaviour and conservation* (1st ed.). Oxford University Press.

Kunze, C., Wölfelschneider, M., & Rölfer, L. (2021). Multiple driver impacts on rocky intertidal systems: The need for an integrated approach. *Frontiers in Marine Science, 8*, 1–13.

Mangel, J. C., Whitty, T., Medina-Vogel, G., et al. (2011). Latitudinal variation in diet and patterns of human interaction in the marine otter. *Marine Mammal Science, 27*, 1–12.

Mangel, J., Alfaro-Shigueto, J., Verdi, R., & Rheingantz, M. L. (2022). *Lontra felina. The IUCN Red List of Threatened Species* 2022: e.T12303A215395045.

Mattern, T., Ellenberg, U., & Luna-Jorquera, G. (2002). A South American otter *Lontra felina* preys upon chicks of the Peruvian diving petrel Pelecanoides garnotii. *Marine Ornithology, 30*, 95–96.

Medina-Vogel, G., Delgado, C., Alvarez, R., & Bartheld, J. L. (2004). Feeding ecology of the marine otter (*Lutra felina*) in a rocky seashore of the South of Chile. *Marine Mammal Science, 20*, 134–144.

Medina-Vogel, G., Boher, F., Flores, G., et al. (2007). Spacing behavior of marine otters (*Lontra felina*) in relation to land refuges and fishery waste in Central Chile. *Journal of Mammalogy, 88*, 487–494.
Meloro, C., & Tamagnini, D. (2022). Macroevolutionary ecomorphology of the Carnivora skull: Adaptations and constraints in the extant species. *Zoological Journal of the Linnean Society, 196*, 1054–1068.
Molina-Maldonado, G., Calvo-Mac, C., & Medina-Vogel, G. (2022). *Evaluación de potenciales endoparásitos en las relaciones tróficas de Lontra felina.* XLI Congreso de Ciencias del Mar, Santiago, Chile.
Newsome, S. D., Tinker, M. T., Gill, V. A., et al. (2015). The interaction of intraspecific competition and habitat on individual diet specialization: A near range-wide examination of sea otters. *Oecologia, 178*, 45–59.
Nuñez-Placencia, P. (2014). *Ecología alimentaria de la nutria marina chungungo (Lontra felina; Molina, 1782) en la Isla Guafo, Región de Los Lagos, Chile.* Tesis de Licenciatura. Universidad Austral de Chile.
Ostfeld, R., Ebensperger, L., Klosterman, L., & Castilla, J. C. (1989). Foraging, activity budget and social behaviour of the South American marine otter *Lutra felina* (Molina, 1782). *National Geographic Research, 5*, 422–438.
Pizarro-Neyra, J. (2014). *Comportamiento de Lontra felina (Molina) en dos ambientes antrópicos de Tacna, sur del Perú.* XIX Congreso Nacional de Biología, Lima, Perú.
Pizarro-Neyra, J. (2022). Interacción entre *Lontra felina* Molina 1782 (Carnivora: Mustelidae) y *Pelecanus thagus* Molina, 1782 (Pelecaniformes: Pelecanidae) en Tacna, Perú. *Mammal Notes, 8*, 268.
Poblete, A., Górski, K., & Moscoso, J. (2019). Estimación de dietas del chungungo *Lontra felina* (Molina, 1782) en dos localidades de la región del Biobío, Chile. *Gayana, 83*, 1–9.
R Core Team. (2022). *R: A language and environment for statistical computing.*
Rau, J., Oyarzún, C., Vilugrón, J., et al. (2021). Registros de mamíferos presentes en el Área Marina Costera Protegida de Múltiples Usos Lafken Mapu Lahual, sur de Chile. *Revista de Biología Marina y Oceanografía, 56*, 167–171.
Rivadeneira, M. M., & Nielsen, S. N. (2022). Deep anthropogenic impacts on benthic marine diversity of the Humboldt current marine ecosystem: Insights from a quaternary fossil baseline. *Frontiers in Marine Science, 9*, 1–12.
Roughgarden, J. (1972). Evolution of niche width. *The American Naturalist, 106*, 683–718.
Rozzi, R., & Torres-Mura, J. C. (1990). Observaciones del chungungo (*Lutra felina*) al sur de la isla Grande de Chiloé: antecedentes para su conservación. *Medio Ambiente, 11*, 24–28.
Ruelas-Cabana, C., Gonzales, F., Arce Merma, A., et al. (2018). *Estudio de la dieta de la nutria marina "Lontra felina" (Molina 1782) en la cuenca de los ríos Colca-Majes y en la laguna Mamacocha, Arequipa, Perú.* Simposio de Investigación y Conservación de las Nutrias de Latinoamérica, XII Congreso de la Sociedad Latinoamericana de Especialistas en Mamíferos Acuáticos.
Russo, L. F., Meloro, C., De Silvestri, M., et al. (2022). Better sturdy or slender? Eurasian otter skull plasticity in response to feeding ecology. *PLoS One, 17*, 1–19.
Sanino, G. P., & Meza, M. I. (2016). Ecología trófica y simpatría de nutrias (*Lontra felina y Lontra provocax*) en la Reserva Añihué, Patagonia Chilena. *Boletín del Museo Nacional de Historia Natural de Chile, 65*, 279–289.
Sepúlveda, M. A., Valenzuela, A. E. J., Pozzi, C., et al. (2021). *Lontra provocax. The IUCN Red List of Threatened Species* 2021: e.T12305A95970485.
Sielfeld, W. (1990). Dieta del chungungo Lutra felina (Molina, 1782) (Mustelidae, Carnivora) en Chile Austral. *Investigaciones en Ciencia y Tecnología, Serie Ciencias del Mar, 1*, 23–29.
Sielfeld, W. (1992). Abundancias relativas de *Lutra felina* (Molina, 1782) y *L. provocax* (Thomas, 1908) en el litoral de Chile austral. *Investigaciones en Ciencia y Tecnología, Serie Ciencias del Mar, 2*, 3–11.

Slauson, K. M., & Zielinski, W. J. (2017). Seasonal specialization in diet of the Humboldt marten (*Martes caurina humboldtensis*) in California and the importance of prey size. *Journal of Mammalogy, 98*, 1697–1708.

Soto, R. (1996). Estructura gremial de un ensamble de depredadores de la zona intermareal rocosa en Chile central. *Investigaciones Marinas, 24*, 97–105.

Spitz, J., Ridoux, V., & Brind'Amour, A. (2014). Let's go beyond taxonomy in diet description: Testing a trait-based approach to prey-predator relationships. *The Journal of Animal Ecology, 83*, 1137–1148.

Timm-Davis, L. L., Dewitt, T. J., & Marshall, C. D. (2015). Divergent skull morphology supports two trophic specializations in otters (Lutrinae). *PLoS One, 10*, 1–18.

Ugarte-Núñez, J. (2021). La nutria marina (*Lontra felina*) en los Andes del sur del Perú. *Ciencia & Desarrollo, 20*, 71–77.

Valqui, J. (2004). *Comportamiento de la nutria marina Lontra felina (Molina, 1782) en un ambiente antrópico, la bahía de Pucusana – Lima, Perú*. Tesis de Licenciatura. Universidad Nacional Agraria La Molina.

Valqui, J. (2012). The marine otter *Lontra felina* (Molina, 1782): A review of its present status and implications for future conservation. *Mammalian Biology, 77*, 75–83.

Vianna, J. A., Ayerdi, P., Medina-Vogel, G., et al. (2010). Phylogeography of the marine otter (*Lontra felina*): Historical and contemporary factors determining its distribution. *The Journal of Heredity, 101*, 676–689.

Villegas, M., Ebensperger, L., & Aron, A. (2001). Prey selectivity of the marine otter, Lontra felina, in nothern Chile. In Libro Resúmenes, Congreso Ciencias del Mar, 2004. Universidad Católica del Norte, Coquimbo, Chile.

Villegas, M., Aron, A., & Ebensperger, L. (2007). The influence of wave exposure on the foraging activity of marine otter *Lontra felina* (Molina, 1782) (Carnivora: Mustelidae) in northern Chile. *Journal of Ethology, 25*, 281–286.

Whitaker, M. R. L., Baker, C. C. M., Salzman, S. M., et al. (2019). Combining stable isotope analysis with DNA metabarcoding improves inferences of trophic ecology. *PLoS One, 14*, 1–15.

Yurkowski, D. J., Ferguson, S. H., Semeniuk, C. A. D., et al. (2016). Spatial and temporal variation of an ice-adapted predator's feeding ecology in a changing Arctic marine ecosystem. *Oecologia, 180*, 631–644.

Chapter 4
Marine Otter Conservation in Peru

Jeffrey C. Mangel, Joanna Alfaro-Shigueto, Clara Ortiz-Alvarez, and Carlos Calvo-Mac

Abstract The marine otter (*Lontra felina*), or chungungo, is one of the three species of otters present in Peru. The species can be found from the northern region of La Libertad (8°04′S) southward into Chile. While the species is primarily marine, its presence inland has now been confirmed up to about 2000 m above sea level. Records of the marine otter in Peru date back to the 1800s but research on the species accelerated after the year 2000, with various Peru-specific and comparative studies published related to species distribution, population size, genetics, and ecology. The species has been categorized in Peru as Endangered since 2004, has been categorized as Endangered since 1996 by the IUCN and is listed in Appendix I by CITES. Current and emerging threats to the species include habitat loss and degradation, fisheries-related threats, health impacts, climate change and other natural threats. Future research opportunities and priorities include investigations of pathogens, microplastic ingestion and contaminants (including through oil spills) as potential health risks, and the impact of habitat loss and degradation and its possible role in isolating or segregating local populations of otters or leading to local extirpations.

J. C. Mangel (✉)
ProDelphinus, Lima, Peru

Centre for Ecology and Conservation, University of Exeter, Cornwall, UK

IUCN Species Survival Commission, Otter Specialist Group, Gland, Switzerland
e-mail: jeff@prodelphinus.org

J. Alfaro-Shigueto
ProDelphinus, Lima, Peru

Centre for Ecology and Conservation, University of Exeter, Cornwall, UK

IUCN Species Survival Commission, Otter Specialist Group, Gland, Switzerland

Carrera de Biologia Marina, Universidad Cientifica del Sur, Lima, Peru
e-mail: jalfaros@cientifica.edu.pe

C. Ortiz-Alvarez
ProDelphinus, Lima, Peru

L. Ayala et al. (eds.), *Marine Otter Conservation*,
https://doi.org/10.1007/978-3-031-53931-2_4

Keywords Threats · Habitat loss · Conservation challenges · Anthropogenic impacts

4.1 Introduction

The marine otter, *Lontra felina*, is one of the three species of otters present in Peru, with its counterparts *Lontra longicaudis* and *Ptenoura brasiliensis* inhabiting freshwater ecosystems. Common names for *Lontra felina* in Peru include nutria marina, nutria de mar, gato marino, chungungo, chingungo, ansumo, and huallaque. Along the Pacific coast of South America, the marine otter can be found from the northern region of La Libertad (8°04′S) in Peru southward into Chile. While the species is primarily marine, its presence inland has now been confirmed up to about 2000 meters above sea level (Ugarte-Núñez, 2021). Records of the marine otter in Peru date back to the 1800s (summarized in Appendix 1 of Apaza & Romero, 2012), with sporadic reports into the 1900s about the species providing limited details about its distribution, abundance, behavior and interactions with human communities (e.g., Mann, 1945; Grimwood, 1968; Moller-Hergt, 1976; Brack Egg, 1978; Brownell, 1978). These early reports describe the species as 'everywhere rare', with an estimated 200–300 animals estimated present along the entire Peru coast in 1968 (IUCN, 1970 as cited by Brownell, 1978). Research on the species accelerated after the year 2000, with various Peru-specific and comparative studies published related to species distribution, population size, genetics, and ecology (e.g., Apaza & Romero, 2012; Ortiz-Alvarez et al., 2021; Mangel et al., 2011; Alfaro-Shigueto et al., 2011; Valqui et al., 2010; Vianna et al., 2010). Nevertheless, research gaps remain which would help inform the development of informed conservation metrics and measures.

In this chapter we review the threats faced by the species, caused primarily by anthropogenic activities including habitat loss, fisheries interactions, among others. We also discuss possible paths forward for marine otter conservation and management at a local and regional scale.

IUCN Species Survival Commission, Otter Specialist Group, Gland, Switzerland

Carrera de Biologia Marina, Universidad Cientifica del Sur, Lima, Peru

Sea Mammal Research Unit, University of St Andrews, St Andrews, UK
e-mail: clara@prodelphinus.org

C. Calvo-Mac
IUCN Species Survival Commission, Otter Specialist Group, Gland, Switzerland

Proyecto Anzumo, ONG ConservAcci6n, Lima, Peru
e-mail: carlos.calvo@upch.pe

4.2 Current and Emerging Threats for Marine Otters in Peru

4.2.1 Habitat Loss and Degradation

Coastal development has a continuing growth, with 83% of the global population living in developing countries (Neumann et al., 2015). The accelerated urbanization of coastal areas and rapid population growth throughout the species distribution in both Peru and Chile has increased pressure on coastal marine ecosystems, causing the loss and fragmentation of marine otter habitat (Ortiz-Alvarez et al., 2021).

Consequences of this habitat degradation related to rapid urban growth includes pollution from the presence of solid waste such as plastics. For marine otters, a recent study identified the presence of microplastics such as fragments, beads and fibers in otter feces collected in Punta Corrientes, located on the central coast of Peru (Santillán et al., 2020). The presence of copper mining activity, mainly in southern Peru, has also been reported as impacting otter habitats (Apaza et al., 2003; Mangel & Alfaro-Shigueto, 2004; Mangel et al., 2011; Valqui, 2012).

In January 2022, an oil spill along the central coast of Peru affected about 150 km^2 within the habitat of marine otters around Lima and Callao. An estimated eight marine otters died as a result of exposure to this spill (SERFOR unpublished information), with some of these animals having been taken to a temporary rehabilitation center where they later died. No necropsies were conducted on any individuals—a missed opportunity to better understand the impacts of oil spills on the species. The continued exposure of spilled oil on intertidal areas over a decade has been a source of exposure to sea otters *Enhydra lutris*, with evidence of persistent oil through the trophic web (Bodkin et al., 2012). This exposure of sea otters to oil is a cause of concern as sub-lethal chronic effect have been documented when compared with non-exposed individuals (Miles et al., 2012). Such potential impacts on marine otters will need to be monitored closely over the coming years.

Other emerging threats related to habitat loss include the construction of infrastructure such as the Terminal Portuario Multiproposito de Chancay and Puerto de Pucusana, areas currently inhabited by marine otters (Valqui, 2012; Ortiz-Alvarez et al., 2021). The construction of dams has also been highlighted as having adverse consequences to otter populations—including habitat loss and fragmentation, alteration in their distribution, and the reduction of foraging areas (Pedroso et al., 2014). In Peru, the future construction of hydroelectric dams in Mamacocha, Arequipa, southern Peru, may represent a threat to riverine areas, and a unique population of marine otters located about 170 miles from the ocean, and 1900 m above sea level first reported around 1997 (Ugarte-Núñez, 2021).

4.2.2 Fur Trade

Unlike in Chile where the fur trade seems to have been a severe threat historically, there is no evidence of extensive hunting of marine otters in Peru. However, commercial hunting at small scale has been reported (Brack Egg, 1978), with more recent reports in Peruvian localities such as Samanco (9°S) (Sánchez & Ayala, 2006), La Libertad (15°S) and Morro Sama (18°S) (Apaza et al. 2003), La Unión (15°S) (Ugarte-Núñez, 2021), and southern Chile (39°S) (Valqui & Rheingantz, 2015).

4.2.3 Fisheries-Related Threats

Interactions of marine otters with fisheries have been reported for decades (Grimwood, 1968; Brownell, 1978; IUCN, 1982). In southern Peru, marine otters have been reported interacting with the changallo shrimp (*Cryphiops caementarius*) fisheries in Arequipa (Brownell, 1978). These early reports indicate that otters were frequently killed because of the damage caused to prawn stocks (Grimwood, 1968; Brownell, 1978). Grimwood (1968) further indicated that marine otters, "are also subject to much casual persecution by fishermen and by owners of firearms, who find them convenient living targets." More recent reports are from the coastal areas of Moquegua, Peru, where these fishery interactions occur as incidental capture or bycatch of otters during fishing operations (Mangel et al., 2011) when gillnets are set in the water and animals are incidentally captured. Another reported threat to marine otters comes from illegal dynamite fishing (Valqui & Rheingantz, 2021).

4.2.4 Health Impacts

One of the main threats to marine otter health is their close proximity to human settlements, as well as the presence of domestic and exotic animals such as dogs, cats, and rats, which can act as reservoirs for a range of pathogens (Medina-Vogel, 2010). Otters are particularly at risk of exposure to zoonotic pathogens in anthropized areas such as ports and docks, where they may come into contact with contaminated water or food (Miller et al., 2002; Burgess et al., 2018). Moreover, the desertic coast of Peru, where the human population is concentrated in coastal cities and towns poses an additional risk to the health of otters due to pollution from mining and agricultural activities that are carried by rivers to their habitat.

A marine otter with a verminous pneumonia, caused by a Metastrongyloidae family, was found in the central coast of Peru (Gonzales-Viera et al., 2011). A study conducted on feces collected in the central coast of Peru identified the presence of bacteria with zoonotic potential causing morbidity and mortality in humans and other aquatic organisms (Calvo-Mac, 2014). Among the genera identified in that study were *Vibrio* sp. and *Aeromonas* sp., with the isolated species of the genus *Vibrio* being resistant to antibiotics such as ampicillin, a synthetic penicillin (Calvo-Mac, 2014). Additionally, an isolation of Vibrio genus was found to be multi-resistant to five antibiotics (Calvo-Mac, 2014). Also, the presence of domestic animals such as dogs and cats can result in negative interactions (e.g., dog attacks) reported for different areas of Peru (Mangel & Alfaro-Shigueto, 2004; Mangel et al., 2011; Pizarro, 2008).

4.2.5 Climate Change

Climate change manifested by ocean warming, acidification, sea level rise and the intensification of severe weather events is expected to produce changes for the marine ecosystem affecting its functionality, resilience and biodiversity (Gissi et al., 2021). Intertidal coastal zones—the habitat of marine otters—will be particularly exposed to extreme weather events which in addition to other climate change manifestations may cause habitat fragmentation or the reduction of suitable areas, as has been shown for freshwater otters (Cianfrani et al., 2018; Cabral et al., 2019). In addition, it is important to consider the synergistic effect of climate change and other human stressors on marine otters. For instance, the increase of precipitation along the coast of Peru could lead to higher levels of contaminants and pathogens in the coastal ecosystem, which could affect the health of marine otters (Miller et al., 2002; VanWormer et al., 2016).

4.2.6 Natural Threats

No natural predators of the marine otter have been documented but the killer whale, *Orcinus orca*, was posited by Cabello (1978) as a possible natural predator. Coastal sharks may also pose a threat to adult individuals while juvenile individuals may be preyed upon by birds of prey (Cabello, 1983). Likewise, the El Niño Southern Oscillation (ENSO) also represents a natural threat that may impact the marine otter population due to the dramatic and cascading ecosystem impacts that can result.

4.3 Future Research and Opportunities

As a top predator in the ecosystem, the health of the marine otter is an important indicator of the overall health of the marine ecosystem it inhabits (Estes, 1989). The presence of enteroparasites such as *Eimeria* sp., *Strongyloides* sp., *Anisakis simplex*, and eggs of Ancylostomidae and Metastrongyloidae have been detected in the feces of marine otters (Vega-Feria et al., 2018). These findings suggest that marine otter feces may serve as indicators of pathogenic contamination in the coastal ecosystem, which warrants further investigation. Despite the potential risk of other pathogens reported in marine otter feces such as vibrio and aeromonas (Calvo-Mac, 2014) few specific studies have been conducted on the health of marine otters and their exposure to these pathogens. Avian influenza H5N1 in sea lions and seabirds in Peru (Leguia et al., 2023), could suggest the need to monitor these pathogens in marine otters as well, as there is a potential for spreading to other marine fauna.

In addition to pathogens, contaminants such as microplastics are also a potential health risk for marine otters. A recent study found microplastics in the feces of otters collected in Punta Corrientes, Peru, highlighting the need for research into the impact of microplastics on the health of the species (Santillán et al., 2020). Furthermore, there is a need to investigate the presence of contaminants from extractive and productive industries such as agriculture and mining, as many of the fish and invertebrates that are prey for marine otters are used as indicators of metal pollution (Loaiza et al., 2022). The oil spill along the northern coast of the Lima region in 2022 was the first time that mortality associated with this type of event was registered (Hooker & Pizarro-Neyra, 2022). However, the long-term consequences on the health of local marine otter populations remain unclear, and it is necessary to evaluate these consequences, which can include impacts on their reproductive and immune systems and ultimately affect the survival of the population. Added to it, future work is needed to have a clear contingency plan for future oil spills and other environmental crisis that could affect the already restricted habitat for marine otters.

To better understand the magnitude of the impact of threats such as climate change and other anthropogenic stressors on marine otter population it is necessary to further study key ecological aspects of marine otters. Improving our understanding of ecological factors related to marine otter occurrence will contribute with the prioritization of extant and emerging risks that can affect the survival of this species. The use of fecal samples as a non-invasive method to monitor the health of marine otters and the ecosystem they inhabit presents an opportunity for future research. The implementation of a coordinated stranding network among various authorities, universities, and civil organizations could also facilitate the monitoring of marine otters, the better understanding of their health and possible causes of death. Given the species reliance on rocky shoreline habitats separated by expanses of sandy beaches, the impact of habitat loss and degradation and its possible role in isolating or segregating local populations of otters or leading to local extirpations (e.g., through oil spills) will also require continued monitoring.

4.4 Conservation Framework

The IUCN (International Union for the Conservation of Nature) global listing for the marine otter began in 1982 with its categorization as Vulnerable. The species categorization was then changed to Endangered in 1996 with the most recent reassessment occurring in 2021 and maintaining the categorization as Endangered

(Valqui & Rheingantz, 2021). The species is also listed on Appendix I of CITES (Convention on International Trade in Endangered Species of Wild Fauna and Flora) and is categorized as Endangered by the Endangered Species Act of the United States.

In Peru, the species has been categorized as Endangered since 2004 (MINAG 2004, and later DS 004-2014-MINAGRI). Although some level of protection may predate this as IUCN (1982) (citing Brack Egg, 1978 and Anon., 1976) refers to a closed season for the species that ran from 15 December to 21 March. And IUCN (1982) indicates that the species was legally protected in Peru since 1977. The establishment of a National Plan of Action for the species would also be a way to help identify, address and mitigate threats to otters. In 2021 the Peruvian government through the Servicio Nacional Forestal y de Fauna Silvestre SERFOR (Ministry of Agriculture) organized a meeting to start discussions of this management document, which could have a positive impact and be an important milestone for the conservation of the species in Peru and help set the agenda for future conservation and monitoring actions. Similar plans are available for other species including jaguars, vicunas, and sea turtles. Some coastal areas, potential habitats for the marine otter, are also part of Peru's system of Natural Protected Areas, however, there is no conservation instrument focused on the species. Suggested conservation measures include the creation of a regional conservation area in southern Peru, the "Morro Sama Coastal Marine Regional Conservation Area, Tacna." This area is recognized as a hotspot for the presence of marine otters (Apaza et al., 2003; Valqui, 2012; Mangel et al., 2011). The creation of this area began in 2003, however, its implementation was halted in 2006. Similarly, a recent study recognizes other areas located along the southern coast of the country, such as Quebrada de Burros, Tres Hermanas Punta Picata and Arquillo, as areas of importance for the otter because they provide adequate conditions for its presence and where conservation strategies should be prioritized (Ortiz-Alvarez et al., 2021).

References

Alfaro-Shigueto, J., Valqui, J., & Mangel, J. C. (2011). New record of the marine otter Lontra felina (Molina, 1782) north to its current distribution. *Ecología Aplicada, 10*(2), 87–91.

Anon. (1976). *Marine Otters*. Scientific Consultation on Marine Mammals. Bergen, Norway, 31 August – 9 September, 1976. ACMRR/MM/SC/\X/G Rep 5.

Apaza, M., & Romero, L. (2012). Distribution and observations on the population of marine otters Lontra felina (Molina 1782) in Peru. *Revista Peruana de Biología, 19*(3), 285–298.

Apaza, M., Valqui, J., & Castañeda, C. (2003). Estado de Conservación de las Poblaciones de Spheniscus humboldti y Lontra felina en la Costa Norte del Perú. In *Instituto Nacional de Recursos Naturales (INRENA)*.

Bodkin, J. L., Ballachey, B. E., Coletti, H. A., Esslinger, G. G., Kloecker, K. A., Rice, S. D., et al. (2012). Long-term effects of the 'Exxon Valdez'oil spill: Sea otter foraging in the intertidal as a pathway of exposure to lingering oil. *Marine Ecology Progress Series, 447*, 273–287.

Brack Egg, A. (1978). Situacion actual de las nutrias (Lutrinae, Mustelidae) en el Peru. In N. Duplaix (Ed.), *Otters: Proceedings of the first working meeting of the otter specialist group* (pp. 76–85). International Union for Conservation of Nature and Natural Resources.

Brownell, R. L., Jr. (1978). Ecology and conservation of the marine otter L. felina. In N. Duplaix (Ed.), *Otters: Proceedings of the first working meeting of the otter specialist group* (pp. 104–106). International Union for Conservation of Nature and Natural Resources.

Burgess, T. L., Tinker, M. T., Miller, M. A., et al. (2018). Defining the risk landscape in the context of pathogen pollution: Toxoplasma gondii in sea otters along the Pacific rim. *Royal Society Open Science, 5*. https://doi.org/10.1098/rsos.171178

Cabello, C. C. (1978). The sea otter L. felina on Chiloé Island. In *Proceedings of the first international otter colloquium, 1*, 118–119.

Cabello, C. C. (1983). La nutria de mar en la Isla de Chiloe. *Boletín Tecnico CONAF,* 37–61.

Cabral, H., Fonseca, V., Sousa, T., & Costa Leal, M. (2019). Synergistic effects of climate change and marine pollution: An overlooked interaction in coastal and estuarine areas. *International Journal of Environmental Research and Public Health, 16*(15), 2737.

Calvo-Mac, C. (2014). *Presencia de bacterias con potencial zoonótico y resistencia antibiótica en heces de nutrias marinas (Lontra felina) en la costa centro del Perú*. Universidad Peruana Cayetano Heredia.

Cianfrani, C., Broennimann, O., Loy, A., & Guisan, A. (2018). More than range exposure: Global otter vulnerability to climate change. *Biological Conservation, 221*, 103–113.

Estes, J. A. (1989). Adaptations for aquatic living by carnivores. In J. L. Gittleman (Ed.), *Carnivore behavior, ecology, and evolution* (pp. 242–282). Springer.

Gissi, E., Manea, E., Mazaris, A. D., Fraschetti, S., Almpanidou, V., Bevilacqua, S., et al. (2021). A review of the combined effects of climate change and other local human stressors on the marine environment. *Science of the Total Environment, 755*, 142564.

Gonzales-Viera, O., Chavera, A., Yaipén-Llanos, C., & Perales-Camacho, R. (2011). Histopathological aspects and etiology of pneumonias in stranded marine mammals from Lima, Peru. *Brazilian Journal of Veterinary Pathology, 4*(1), 23–29.

Grimwood, I. R. (1968). *Notes on the distribution and status of some Peruvian mammals* (Special publication no. 21) (86 pp). American Committee for International Wild Life Protection and New York Zoological Society.

Hooker, Y., & Pizarro-Neyra, J. (2022). *Evaluación rápida del estado de las poblaciones de nutria marina Lontra felina en el sector de mayor afectación por el derrame de petróleo de La Pampilla*. Lima.

IUCN. (1970). *Marine otter: Chingungo Lutra Felina, Code No. MA/99/Lutra Fel, Red Data Book.* Survival Service Commission, International Union for the Conservation of Nature and Natural Resources.

IUCN. (1982). Marine otter. In *The IUCN mammal red data book: Part 1 - Threateneed mammalian taxa of the Americas and the Australasian zoogeographic region (excluding cetacea)* (Compiled by J. Thornback and M. Jenkins) (pp. 353, 516 pp–356). World Wildlife Fund and United Nations Environment Programme.

Leguia, M., Garcia-Glaessner, A., Munoz-Saavedra, B., Juarez, D., Barrera, P., Calvo-Mac, C., Jara, J., Silva, W., Ploog, K., Amaro, L., Colchao-CLaux, P., Johnson, C. K., Uhart, M. M., Nelson, M. I., & Lescano, J. (2023). Highly pathogenic avaian influenza A (H5N1) in marine mammals and seabirds in Peru. *Nature Communications*. https://doi.org/10.1038/s41467-023-41182-0

Loaiza, I., De Boeck, G., & De Troch, M. (2022). Peruvian marine ecosystems under metalcontamination: First insights for marine species consumption and sustainable management. *Science of the Total Environment, 826*, 154132. https://doi.org/10.1016/J.SCITOTENV.2022.154132

Mangel, J. C., & Alfaro-Shigueto, J. (2004). Comunidades pesqueras y la conservacion de la nutria marina (Lontra felina) en el sur de Peru. *Revista de Conservación Regional, 2*, 5–10.

Mangel, J. C., Whitty, T., Medina-Vogel, G., Alfaro-Shigueto, J., Cáceres, C., & Godley, B. J. (2011). Latitudinal variation in diet and patterns of human interaction in the marine otter. *Marine Mammal Science, 27*(2), E14–E25. https://doi.org/10.1111/j.1748-7692.2010.00414.x

Mann, G. (1945). Mamiferos de Tarapaca. *Biologica Santiago, 2*, 23–134.

Medina-Vogel, G. (2010). Ecology of emerging infectious diseases and wild species conservation. *Archivos de Medicina Veterinaria, 42*, 11–24. https://doi.org/10.4067/s0301-732x2010000100003
Miles, A. K., Bowen, L., Ballachey, B., Bodkin, J. L., Murray, M., Estes, J. L., et al. (2012). Variations of transcript profiles between sea otters Enhydra lutris from Prince William Sound, Alaska, and clinically normal reference otters. *Marine Ecology Progress Series, 451*, 201–212.
Miller, M., Gardner, I., Kreuder, C., et al. (2002). Coastal freshwater runoff is a risk factor for *toxoplasma gondii* infection of southern sea otters (*Enhydra lutris nereis*). *International Journal for Parasitology, 32*, 997–1006. https://doi.org/10.1016/S0020-7519(02)00069-3
Moller-Hergt, S. (1976). *Analisis de la situacion actual de los mamiferos, aves y reptiles silvestres del Peru* (177 pp.). Ministerio de Agricultra, Direccion General Forestal y de Fauna.
Neumann, B., Vafeidis, A. T., Zimmermann, J., & Nicholls, R. J. (2015). Future coastal population growth and exposure to sea-level rise and coastal flooding – A global assessment. *PLoS One, 10*(3), e0118571. https://doi.org/10.1371/journal.pone.0118571
Ortiz-Alvarez, C., Alfaro-Shigueto, J., & Mangel, J. C. (2021). Insights into marine otter (Lontra felina) distribution along the Peru coastline. *Marine Mammal Science, 37*(3), 993–1007.
Pedroso, N. M., Marques, T. A., & Santos-Reis, M. (2014). The response of otters to environmental changes imposed by the construction of large dams. *Aquatic Conservation: Marine and Freshwater Ecosystems, 24*(1), 66–80.
Pizarro, J. (2008). Mortality of the marine otter (Lontra felina) in southern Peru. *IUCN Otter Specialist Group Bulletin, 25*(2), 94–99.
Sánchez, R., & Ayala, L. (2006). *Evaluacion de la distribucion y abundancia del gato marino Lontra felina en la costa central y norte del Peru durante el otono del 2006 Reporte APECO*, 33p.
Santillán, L., Saldaña-Serrano, M., & De-La-Torre, G. E. (2020). First record of microplastics in the endangered marine otter (Lontra felina). *Mastozoología Neotropical, 27*, 211–215. https://doi.org/10.31687/saremMN.20.27.1.0.12
Ugarte-Núñez, J.A. (2021). La nutria marina (*Lontra felina*) en los Andes del sur del Peru. *Cienc Desarro, 20*,71–77. https://doi.org/10.33326/26176033.2021.1.1109.
Valqui, J. (2012). The marine otter Lontra felina (Molina, 1782): A review of its present status and implications for future conservation. *Mammalian Biology, 77*, 75–83. https://doi.org/10.1016/j.mambio.2011.08.004
Valqui, J., & Rheingantz, M. L. (2015). Lontra felina. In *The IUCN red list of threatened species 2015: e.T12303A21937779*. https://doi.org/10.2305/IUCN.UK.2015-2.RLTS.T12303A21937779.en
Valqui, J., & Rheingantz, M. L. (2021). Lontra felina. In *The IUCN red list of threatened species 2021: e.T12303A95970132*. https://doi.org/10.2305/IUCN.UK.2021-3.RLTS.T12303A95970132.en
Valqui, J., Hartl, G. B., & Zachos, F. E. (2010). Non-invasive genetic analysis reveals high levels of mtDNA variability in the endangered South-American marine otter (Lontra felina). *Conservation Genetics, 11*(5), 2067–2072. https://doi.org/10.1007/s10592-010-0069-9
VanWormer, E., Carpenter, T. E., Singh, P., et al. (2016). Coastal development and precipitation drive pathogen flow from land to sea: Evidence from a toxoplasma gondii and felid host system. *Scientific Reports, 6*, 1–9. https://doi.org/10.1038/srep29252
Vega-Feria, K., Calvo-Mac, C., Ortiz-Alvarez, C., & Serrano-Martínez, E. (2018). Endoparásitos en heces de nutria marina (Lontra felina) en la costa de Perú. In: *XII Congreso de la Sociedad Latinoamericana de Especialistas en Mamíferos Acuáticos – RT 18*, p 335.
Vianna, J. A., Ayerdi, P., Medina-Vogel, G., Mangel, J. C., Zeballos, H., Apaza, M., & Faugeron, S. (2010). Phylogeography of the marine otter (Lontra felina): Historical and contemporary factors determining its distribution. *Journal of Heredity, 101*(6), 676–689. https://doi.org/10.1093/jhered/esq088

Chapter 5
Conserving Marine Otter *Lontra felina* in the One of Oldest Marine Protected Areas of Southeast Pacific

Raúl Sánchez-Scaglioni, Liliana Ayala, and Patricia Saravia-Guevara

Abstract Marine otter was one of the reasons to create Paracas National Reserve (PNR), the first protected marine and coastal area in Peru. This natural protected area included the marine otter in its management plans, including the monitoring of the population and identifying a minimal number of otters. In this chapter, we gather and analyze previous studies that assess the marine otter population inside the PNR, including monitoring developed by governmental officials. In addition, we identify the actions around marine otters were included in the management plans in the PNR. We found differences between the methods of censuses of marine otters by different researchers and the PNR officials. We recommend a unique methodology with the agreement of the specialist to report the number of otters, daytime, season, length of assessed area, time used for the census, among others, and considering the logistical constraints. The marine otter is an important conservation object in the PNR management plans. All management plans from 1979 to 2020, include actions to protect marine otters. An effective management of the PNR and coordination with other governmental sectors should be implemented to promote marine otter conservation.

Keywords Paracas · Marine otter · Protected area · Management plan · Conservation

---

R. Sánchez-Scaglioni
Asociación Peruana para la Conservación de la Naturaleza, Lima, Peru
e-mail: rsanchezscaglioni@apeco.org.pe

L. Ayala (✉)
Carrera de Biología Marina, Universidad Científica del Sur, Lima, Peru

Asociación Peruana para la Conservación de la Naturaleza, Lima, Peru
e-mail: layalaa@cientifica.edu.pe

P. Saravia-Guevara
Reserva Nacional de Paracas, Servicio Nacional de Áreas Naturales Protegidas por el Estado, Paracas, Peru
e-mail: psaravia@sernanp.gob.pe

L. Ayala et al. (eds.), *Marine Otter Conservation*,
https://doi.org/10.1007/978-3-031-53931-2_5

5.1 Introduction

Protected areas are considered an essential strategy for habitat and species conservation, as they contain more biodiversity related to their areas than unprotected sites (Geldmann et al., 2013; Gray et al., 2016). In Peru, the objective of the system of protected areas is to conserve its biological diversity (Republica del Perú, 1997). Along the Peruvian coast, there are protected areas of different categories with different levels of restrictions regarding the activities allowed inside them: National Sanctuaries, National Reserves, Landscape Reserves, Wildlife Refuge, Reserved Zones, Regional Conservation Areas and Private Conservation Areas. Natural protected areas are managed by the Servicio Nacional de Areas Naturales Protegidas por el Estado (SERNANP National Service of Natural Areas Protected by the State). However, at least for terrestrial species, these protected areas do not provide sufficient protection for many species according to the specified conservation goals, since species on the coast are the worst represented in the current protected area system (Fajardo et al., 2014).

Some studies have reported the distribution and abundance of the marine otter along the Peruvian coast in the last decades (Valqui, 2012; Apaza & Romero, 2012; Ortiz-Alvarez et al., 2021). Marine otter has a discontinuous distribution following rocky shore patches, caves, or rock cracks above the high tide line (Medina-Vogel et al., 2008). In addition, this discontinuous distribution can be affected by human constructions that marine otter use as habitat (Cursach et al., 2012). Then, marine-coastal protected areas containing rocky seashore patches could provide good habitat conditions for marine otters, ensuring the survival of the species (Valqui, 2012; Ortiz-Alvarez et al., 2021).

Marine otter is a species that has been reported in six of the marine coastal protected areas in Peru: Paracas National Reserve (PNR), Guano Islands, Islets, and Capes National Reserve System of Peru, Ancon Reserved Zone, San Fernando National Reserve, Lomas de Atiquipa Private Conservation Area and Santuario Nacional de Mejía (INRENA, 1997; Apaza & Romero, 2012; SERNANP, 2019, 2020). PNR in the central coast of Peru was the first marine-coastal protected area established in Peru in 1975. Only Galapagos National Park is older than PNR in the Southeast Pacific (FAO, 2012). The extraction of natural resources by local people according to a Management Plan is allowed in a National Reserve (República del Perú, 1997).

PNR has an area of 3350 km^2; of which 2176 km^2 cover the marine environment including islands and islets, 1174 km^2 correspond to coastal desert (SERNANP, 2016). Parts of PNR are the guano islands La Vieja and Santa Rosa formerly managed by the Compañia Administradora del Guano and currently managed by AGRORURAL, an organ part of the Ministry of Agriculture. PNR was the only area conserving a representative portion of the Peruvian marine environment until 2009. PNR is in one of the most biological productive marine areas in the world, serving as home of 300 fish species, over 200 migratory bird species (60 of which migrate between Peru and the United States), and marine mammals and reptiles (Millennium

Ecosystem Assessment, 2005). In addition, there is a high biomass of benthic organisms such as mollusks and crustaceans in shallow waters and under normal upwelling conditions (Taylor et al., 2008). However, events of coastal hypoxia and anoxia currently characterize Paracas Bay (north of PNR) during summer and autumn, favored by oceanography, topography, river discharge and anthropogenic activities (Pitcher et al., 2021).

The marine otter was always a conspicuous species in this protected area, but it began to be monitored in the late 1990s, and the species is considered an indicator of what is defined as a priority element to conserve within PNR. In this case, the ecosystems of islands, islets, points, and cliffs, important as they are breeding, feeding and resting areas for threatened wildlife (SERNANP, 2016). Analyses of the presence and conservation status of the marine otter in the older marine-coastal protected area in Peru would give us a panorama of the status of that threatened species. The trends of that species have not been modeled but data since 1970s showed the increase of marine otters in the Peruvian coast (Apaza & Romero, 2012).

This study includes a review of published and unpublished literature about distribution and occurrence of marine otter in Paracas National Reserve (PNR). Therefore, we identify the actions around marine otters were included in the management plans in the PNR.

5.2 Methods

5.2.1 Data Collection

Information was solicited to Servicio Nacional de Areas Naturales Protegidas por el Estado (SERNANP National Service of Natural Areas Protected by the State) and Agrorural (Programa de Desarrollo Productivo Agrario Rural-Development Program of Agricultural Products in Rural Area) through the protocol established by Law No. 27806 on access to public information. We solicited information about marine otter population assessment of last 25 years, particularly marine otter numbers and localities where they were observed and registered.

In addition, the management plans of PNR since its creation (1975) were reviewed. We identified the actions around marine otters in the plans.

5.2.2 Literature Review

A review for published and unpublished literature was made about occurrence of marine otter in PNR. The sources were Google Scholar, ResearchGate, Science Direct, JSTOR and digital repositories during the period 1986–2020. The keywords and phrases in English and Spanish searched were "marine otter", "marine otters in Peru", "*Lontra felina*", "Paracas National Reserve". In addition, the authors had unpublished information. A total of seven papers were collected, where four papers were studies in PNR and three were studies in the Peruvian coast (including PNR). Methods, localities, numbers, seasons, type of habitat of each locality were reported.

5.3 Results

5.3.1 First Censuses of Marine Otter in Paracas National Reserve

During summer and winter of 1989, marine otters were counted between 7 and 19 h in San Gallan Island, only on the accessible part of the shore (about 6 km), the inaccessible part corresponds to cliffs and reefs that do not allow landing on the shore (Sanchez-Scaglioni, 1990). San Gallan I. is part of PNR, with an area of 9.32 km^2, 14 km shore length, width of 4 km and 6 km offshore. It is mostly dry, barren, and dusty, but with high hills reaching well into the clouds, and only there, in the moist altitudes, teeming with plant life (Coker, 1919). On this island there are no people permanently, except occasionally fishermen or shellfish extractors as well as wildlife researchers and surfers.

The second census covers the island and peninsula of PNR during 1991 (Sanchez-Scaglioni, 1992). Marine otters were observed from vessel and land. He observed marine otters directly or look for feces or burrows. He stayed at least 3 h in the locality when he finds marine otters or its trace. He uses binoculars and observes marine otters until 100 m. Numbers of marine otter from both studies are presented in Table 5.1.

Table 5.1 First censuses reported of marine otter in Paracas National Reserve during 1989 and 1991

Year	Season	Report	Localities	Individuals San Gallan I.	Total individuals	Otters/km San Gallan I.
1989	Summer and winter	Direct sighting	1	9	9	1.5
1991	Winter	Direct sighting	7	13	25	2.2

Sanchez-Scaglioni (1990, 1992)

5.3.2 PNR Data of Marine Otters

PNR through SERNANP provided information from 2002 to 2020. Agrorural provided counts from 2002 to 2022 in La Vieja and Santa Rosa Islands as part of PNR. In those islands, Agrorural as a governmental organism gather and later commercialize guano. Another important island in PNR is San Gallan although is not a guano island with no presence of Agrorural guards. As part of conservation of marine fauna in guano islands, Agrorural island guards report the population status of guano birds and other species like marine otters. La Vieja I. is 1.6 km from Santa Rosa I. with 11 km^2 and 390 m of high. La Vieja had less guano birds than Santa Rosa (SERNANP, 2016, Valqui et al. 2010).

PNR observers counted marine otters from 9 to 16 h. Observers stayed 1 to 2 h in the observation sites which were previously established. Observers use binoculars and report direct sightings until 250 m.

5.3.3 Marine Otter on Other Islands of the PNR

The first and second day of every month, island guards count marine otters using binoculars 10 × 50 (Agrorural, 2023). Two marine otters were reported only in January 2022 on La Vieja Island. Meanwhile, in Santa Rosa Island, marine otters were reported between 2015 and 2022 (3.8 ± 2.7; 1–6, $n = 5$). In Table 5.2, otter/km was not calculated in La Vieja I. because in the same period, island guards count seabirds, sea lions in part of the 17 km of perimeter of that island.

5.3.4 Comparison Between Otter Counts by PNR Officers and Counts Conducted by Other Researchers

Four studies have information of marine otters in the same years of PNR. Two authors (Apaza et al., 2004; Sanchez-Scaglioni, 2004) reported direct observations, then feces and prints; corpses were not considered. Sanchez-Scaglioni (2004)

Table 5.2 Marine otters reported in Santa Rosa and La Vieja islands

Year	Season	Locality	Total individuals	Otters/km
2015	Summer	Santa Rosa I.	1	0.2
2019	Spring	Santa Rosa I.	1	0.2
2020	Winter	Santa Rosa I.	1	0.2
2021	Winter-spring	Santa Rosa I.	3–6	1.4
2022	Summer	Santa Rosa I.	3–6	1.4
2022	Summer	La Vieja I.	2	?

Data provided by Agrorural

Table 5.3 Number of marine otters in Paracas National Reserve for 3 years according to RNP staff and other authors

Year	Author	Season	Method	Localities	Individuals San Gallan I.	Total individuals
2002	Apaza et al. (2004)	Summer	Direct sighting	5	2	11
	PNR	Not reported	Direct sighting	6	ND	5
2004	Sanchez (2004)	Winter	Direct sighting	17	2	28
	PNR	Not reported	Direct sighting	2	ND	2
2008	Valqui et al. (2010)	Winter	Tissues sample collected	12	ND	14
	Ormeño and Anchante (2009)	Summer	Direct sighting	13	ND	35
	PNR	Not reported	Direct sighting	9	ND	17

ND no data

reported three periods to count marine otters: between 11 and 14 h, from 16 h to sunset and from sunrise to 8 h, from major to minor frequency respectively. Apaza et al. (2004) indicated that counts were made during the afternoon. Meanwhile, Valqui (2012) registered prints and collected feces to determine the genetic status of the population of the marine otter. Ormeño and Anchante (2009) reported counts of marine otter for 40 minutes by each locality.

Most of the counts were made by direct observation (83%). The numbers of localities assessed in PNR varied annually (Table 5.3).

5.3.5 *Historical Localities of Marine Otters in the PNR*

A total of 38 sites have been reported by seven studies and the monitoring in PNR and its islands (Sanchez-Scaglioni, 1990, 1992, 2004; Ormeño & Anchante, 2009; Valqui, 2012; Apaza & Romero, 2012; Ortiz-Alvarez et al., 2021) (Table 5.4).

The 44% of sites are in the north, 49% in the south and 7% is for San Gallan, Santa Rosa and La Vieja I. Arquillo is the most assessed site (88% of studies) followed by Yumaque (65%). They are followed by San Gallán, Punta Prieta, Lechuza, La Esperanza, Lagunillas, Santa Maria, La Catedral, Sacasemita (50% each one); La Mina, Erizal (38% each one); Supay, Mendieta, La Zorra, El Queso, El Negro (25% each one) (Fig. 5.1).

Most of the habitat where marine otter was observed in rocks (86%), followed by sand beaches (11%) and a dock (3%). Marine otters were reported in five of seven zoning categories. Human activities reported for 2020 in marine otter localities were nine such as: Macroalgae extraction (45%), vehicle traffic, tourist activities,

Table 5.4 Marine otters reported in 38 locations in PNR

N	Site	Longitude	Latitude	% Occurrence	Min	Max	Number of counts
Sites assessed by governmental officers and researchers							
1	El Negro	76.02	14.44	1.0	3	3	4
2	Antana	76.05	14.41	1.4	1	1	4
3	Barlovento	76.10	14.37	1.4	1	1	5
4	Ventosa	76.11	14.32	1.0	1	1	3
5	Queso	76.13	14.26	4.5	1	1	14
6	La Vieja	76.19	14.27	0.3	2	2	1
7	Santa Rosa I.	76.19	14.27	1.4	1	6	4
8	Los Canastones	76.19	14.18	1.0	1	1	3
9	Caclio	76.20	14.17	1.0	1	1	3
10	Tres Puertas	76.21	14.17	3.5	1	1	10
11	El Chucho	76.23	14.15	3.5	1	1	10
12	Punta Carreta	76.27	14.20	2.4	1	1	7
13	Los Frailes	76.27	13.97	1.0	1	1	3
14	Gallinazo Norte	76.28	14.08	3.5	1	1	11
15	Sacasemita	76.28	14.15	4.5	1	2	14
16	Erizal	76.28	14.12	4.9	1	3	15
17	La Catedral	76.29	13.94	4.5	1	2	15
18	Cueva de la Zorra	76.29	14.08	4.5	1	1	13
19	Balcones	76.29	13.91	5.6	1	5	16
20	Santa Maria	76.30	13.90	5.2	1	5	17
21	Playa Roja	76.30	13.89	3.5	1	1	10
22	La Cruz	76.31	13.79	1.0	1	1	3
23	Lagunilla	76.31	13.90	5.2	1	3	17
24	Esperanza	76.32	13.91	4.5	1	3	15
25	La Mina	76.32	13.91	4.5	1	3	13
26	Punta Prieta	76.33	13.92	5.9	1	3	17
27	Pta. Arquillo	76.35	13.92	5.6	1	7	20
28	Lechuza Baja	76.37	13.91	4.9	1	2	16
29	Mirador de lobos	76.40	13.90	2.8	1	3	10
Other sites assessed by researchers							
30	Morro Quemado	76.13	14.36	0.3	1	1	1
31	Ancla	76.26	14.17	0.3	1	1	1
32	Mendieta	76.26	14.06	0.3	1	1	2
33	Supay	76.28	13.94	0.3	1	2	2
34	Yumaque	76.29	13.91	1.0	1	2	3
35	Playa Talpo	76.34	13.80	0.3	2	2	1
36	Culebras	76.39	13.84	0.7	1	1	1
37	Los Choros	76.40	13.90	0.3	1	1	1
38	San Gallan I.	76.43	13.84	1.4	2	13	4
	Number of marine otters				43	90	

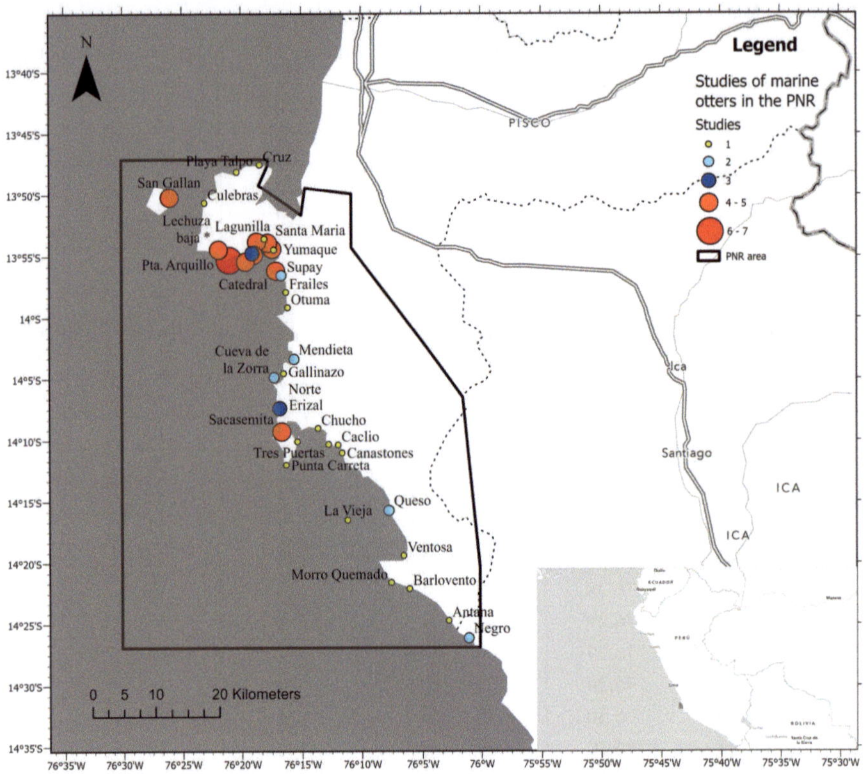

Fig. 5.1 Number of studies to assess marine otters in the Paracas National Reserve in 35 localities according to nine sources

pollution, use of explosive (9% each one), fishing, salt extraction, aquaculture, and aquatic sports (4.5% each one) (Fig. 5.2).

5.3.6 Management Plans

Four management plans were reviewed for the period 1979–2020 (Ministerio de Agricultura, 1979; INRENA, 1996, 2002; SERNANP, 2016). Each plan establishes some actions on marine otter, and they are described in the following paragraphs:

Management Plan, 1979: One of the six objectives of PNR establishes to protect threatened species in the area such as marine otter and other eight species.

Management Plan, 1996: Some research projects are proposed in this plan such as the Inventory of Wildlife Flora and Fauna. Marine otter is included in the inventory with other marine mammals. Other project is studying and monitoring threatened species, marine otter is one of three of them. This study proposed: (i)

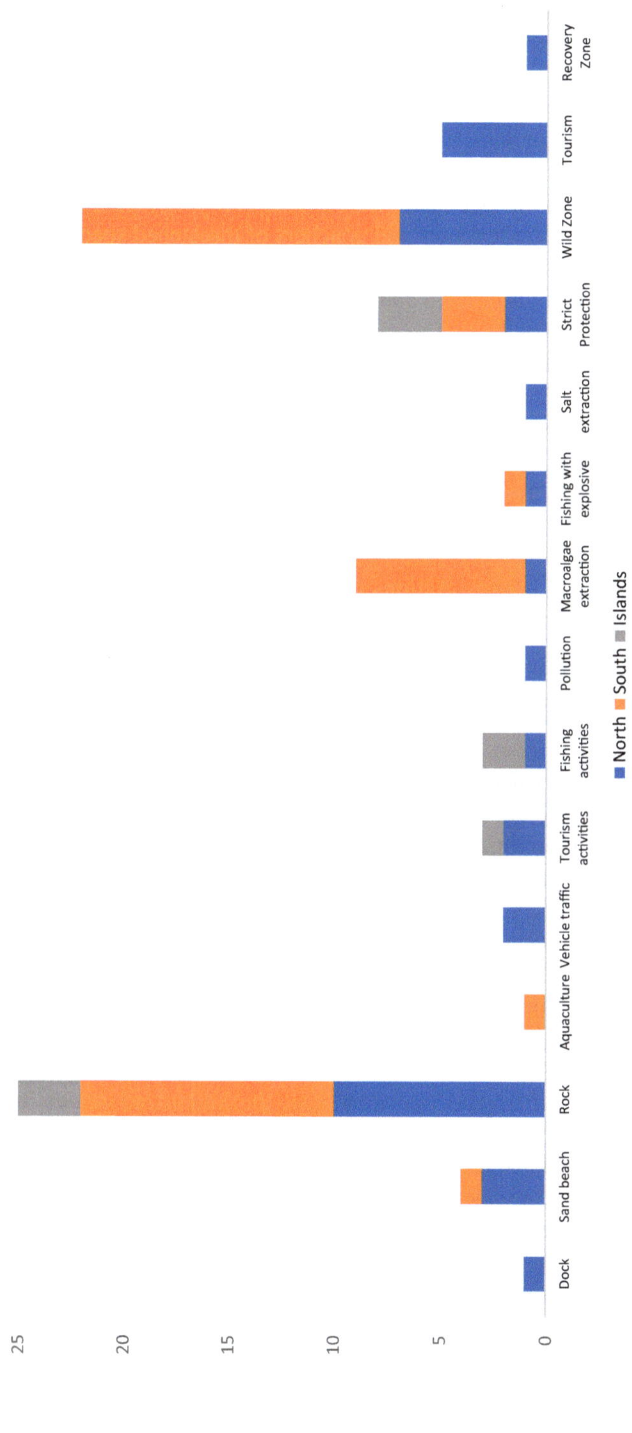

Fig. 5.2 Number of historical localities of marine otter in PNR according to eight sources. Type of habitat, human activities and zones according to management plan of PNR in north, south and islands

Assess population of threatened species, (ii) Management of recovered populations, and (iii) Recovery of the habitats of these species. Budget is calculated, although it seems to be only proposals.

Management Plan 2003–2007: As part of the strategies, the plan establishes to monitor terrestrial biodiversity to know spatio-temporal variability and preventing or controlling threats or loss of functionality. Thus, monitoring plan should be focused on species, natural formations, terrestrial environments or unique landscapes of PNR. The management plan outlined activities including to characterize the illegal trade, trade routes and poachers of threatened species (such as the marine otter). As part of the strategy to increase and strengthen presence and capacity to take actions of local authorities to restore principle of authority.

Management Plan, 2016–2020: The plan outlines that on the islands, islets, points and cliffs of the PNR, there is a base population of between 20 and 40 individuals of marine otters by 2015. Thus, a conservation goal was established, marine otter population should be more than 20 individuals in those places. The Management Plan includes a strategy to implement its environmental aspects, such as monitoring marine otters by four park rangers. Also, the plan acknowledges that marine otter is affected by human perturbation. This perturbation is originated by collecting macroalgae illegally, landing by non-authorized aquatic sportsman and guano extraction in islands. Actions proposed are management of collection of macroalgae and management of touristic activities.

In addition, management plan zoning the PNR in eight categories. Five zones where marine otters were found have the following definitions (Republica del Perú, 1997):

Strict Protection Zone (SP): Few or not intervened ecosystems with fragile or rare species. Only allowed activities such as monitoring, surveillance, and control. In addition, research activities must have authorization from chief of PNR. Vehicles and people are not allowed.

Wild Zone (WZ): Zones less vulnerable than strict protection zones. Activities allowed are monitoring, surveillance, control, sustainable tourism, and environment education. Also, research activities must have authorization from chief of PNR. Areas with little or no intervention human and where the wildlife predominates; but this zone is less vulnerable than Strict Protection Zone.

Touristic and Recreational Use (T): Recreational use compatible with objectives of protected area. Tourists and operators visit the authorized routes, trails, and zones. Artisanal fishery is allowed, aquatic sports are not allowed. In this zone is allowed educational and research activities, as well as infrastructure for services necessary for the access, stay and enjoyment of the visitors, including carriage access routes, shelters and use of vehicles.

Special Use Zone (SU): Zones with human settlements prior to the establishment of protected area. It is allowed the commercial exploitation of approved resources by multisectoral commission.

Recovery Zone (RZ): Temporal zones due to human or nature perturbation have suffered damages. It requires special management to recover quality and environmental stability. In PNR, a dock is used to land resources. The liquid and solid waste must be disposed out of the PNR.

The other zones are Historical-Cultural Zone (HC) and Zone of Direct Use (DU).

5.4 Discussion

The conservation of marine otter population was decisive for the creation of the PNR. Although the reserve has few studies about the status of the marine otter population, the PNR is the only coastal area of Peru practicing a constant effort to monitor and protect this threatened species. Even regardless of the changes in policies, visions, managers, and circumstances that have occurred in the evolution of the management of this natural area protected by the Peruvian government.

There are several differences between the characteristics of censuses of marine otters by three researchers and the PNR, such as number of localities, seasons, even the hours of the censuses. A first recommendation would be a unique methodology with the agreement of the specialist to report the number of otters, daytime, season, length of assessed area, time used for the census, among others, and considering the logistical constraints.

The first report of a marine otter in San Gallan I. was in 1908 (Coker, 1919). Several decades later, the PNR was established in 1975. The first censuses of marine otters reported more than a dozen of marine otters in that island (Sanchez-Scaglioni, 1990, 1992). It was reported one individual each 100 m of shore (Reyes, 1992), although this last report is possible that it refers to the density of individuals in the specific areas observed, not to the entire shore of the island. However, assessed that island implies logistical requirements and adequate oceanographic conditions to be able to disembark. That would be a reason due to PNR did not assess marine otters in San Gallan I. However, an intense assessment in that island is recommended to know the real state of marine otter in the PNR since the island is a refuge to marine otters with no human disturbance by guano harvest, human inhabitants, or touristic activities. San Gallan I. is visited only by researchers, surfers, and shellfish fishers (Catenazzi & Donnelly, 2008). In Santa Rosa Island under isolation, cessation of guano harvest and favorable environmental and food conditions, Humboldt penguins can increase establishing important colonies (Zavalaga & Alfaro-Shigueto, 2015). Thus, San Gallan (Fig. 5.3) and Santa Rosa I. with no human perturbation can be important sites to marine otters in the PNR.

Also, there were historical sites for marine otters most assessed by researchers in the PNR. The first two sites, Arquillo and Yumaque are part of a touristic route and for that reason they are accessible, in this case to researchers (SERNANP, 2016). In addition, Arquillo (Fig. 5.4) was identified as an important area for marine otter due to provide good habitat conditions for marine otters (Ortiz-Alvarez et al., 2021). San Gallan I. is other site most assessed due to the high numbers reported in the first censuses, but the island is not part of the PNR assessments or other most recent

Fig. 5.3 Marine otter in the San Gallan Island. (Photo: Patricia Saravia-Guevara)

Fig. 5.4 Arquillo is an important area for marine otter due to provide good habitat conditions for marine otters in Peru. (Photo: Patricia Saravia-Guevara)

studies. Thus, a logistical issue could limit the sites and the area that could be assessed. Most of the sites assessed are rocky habitats that provide breeding and resting places (Valqui, 2012). In addition, Lagunillas with a human construction, a wharf, are part of the marine otter habitat (Alfaro-Shigueto et al., 2011; Valqui, 2012) (Fig. 5.5). Marine otters forage on the remains of fish discarded by fishermen in Lagunillas (N. de Paz, pers. comm.). However, it has been reported marine otters

Fig. 5.5 Marine otter with two pups in Lagunillas. (Photo: Patricia Saravia-Guevara)

attacked by dogs, poisoned and incidental caught due to that synanthropic behavior in other sites (Mangel et al., 2011; Cursach et al., 2012). Dogs are not allowed in the PNR, although perturbation produced by dogs on wildlife have been reported in the PNR (Huayanca et al., 2017).

Some researchers have dedicated effort to assess marine otter especially in the northern PNR. In that zone, especially Lagunillas Bay, it has been reported DDT concentrations in the mussels *Semimytilus algosus* in Lagunillas Bay, although less than FDA standards (Cabello & Sánchez, 2006). In addition, mollusks have been reported as a prey of the marine otter diet (Biffi & Iannacone, 2010; Poblete et al., 2019). Meanwhile, PNR staff have assessed the southern of PNR, reporting new localities for marine otter in Independencia Bay and southern. Moreover, localities in northern Independencia Bay (14% of localities) are influenced by interactions between wind and sea circulation which stimulate the upwelling process (Quispe et al., 2010).

We are not able to find any reports of otter in Bahia de Paracas, although in the northern of the bay we observe rocky shore, cliffs and small caves that could be used by otters. Bahia de Paracas is characterized by periodic events of anoxia and hypoxia and by the occurrence of algal blooms due to seasonal oceanographic changes and organic matter from the Pisco River which is in the north of the bay (Pitcher et al., 2021). Moreover, an algal bloom caused by *Alexandrium ostenfeldii* and related to the detection of paralytic shellfish toxins has recently been reported (Cuellar-Martinez et al., 2021).

The marine otter is an important conservation object in the PNR management plans. All management plans from 1979 to 2020, include actions to protect marine otters. Even, the first plan mentions the protection of marine otter as an objective of the PNR. Meanwhile, in the second plan marine otter is considered as a research object of two proposals. The third plan mandated that a monitoring plan should be developed and implemented for mammals and proposed to identify persons and their actions to trade threatened species. The last one establishes a minimum size of otter population and actions to face human activities affecting marine otters. The presence of marine otter in management plans allowed to maintain the monitoring during last two decades of this important conservation objective of the PNR. A similar situation is observed in protected areas of Chile, for example, a protected area important to marine otters, Isla Cachagua National Monument established as an objective of its creation: to protect marine species breeding or living in the island (CONAF, 2008). PNR is part of SERNANP (National Service of Natural Areas Protected by the State), which is a specialized public technical agency under the Ministry of Environment of Peru (Ministerio del Ambiente—MINAM). Also, marine otter is protected by Peruvian Government and Agricultural Ministry oversees threatened species (Ministerio de Agricultura, 2006).

Tourism and fishing were activities reported in PNR. Only 15% of assessed localities in PNR are touristic and other areas can be visited by fishers catching with a line and a hook called "pinteros". Then both activities can affect marine otters. Neotropical otters can be affected by tourist activities, changing strategies of forage such as changing times to search food due disturbance (Briones-Salas et al., 2013). However, noninvasive disturbance such as ecotourism may promote giant otter tolerance to the presence of humans (Barocas et al., 2022). In addition, the main causes of death in marine otters at south Peru were related to interaction with the local people and fishing activities (Pizarro Neyra, 2008). Moreover, human perturbance by fishing can produce otters avoid humans, thus, extractive activities may be perceived as riskier (Barocas et al., 2022).

Conservation and monitoring of marine otter is a priority for the PNR. The effort has been concentrated to monitor marine otter population in PNR. Logistical difficulties make it nearly impossible to survey the complete extent of the reserve and contribute to a partial assessment of trends within the PNR. However, if the numbers of marine otter should be maintained, also actions to conserve this population must be faced. These actions have been identified in the last plan, but the most difficult step is to implement their management to promote marine otter conservation activities such as macroalgae collection, tourism, and fishing. A huge task faced for four decades and that require a coordinated work between Environmental, Produce and Tourism ministries for several decades more.

Acknowledgments In memory of biologist Gladys Rojas Dulanto, the first head of the Paracas National Reserve and the first woman designated head of a state-protected natural area in Peru, who died in the exercise of her duty. Also, we would like to thank Paracas National Reserve (as part of SERNANP) and Agrorural which provided information about the numbers of marine otters and their localities from 2002 to 2020. We are particularly grateful to the park rangers who over the years have carried out the monitoring and surveillance work in the RNP Elver, Coronado Flores, Santiago, Herrera Sauñe, Víctor Quispe Huamán, Sérvulo Villagómez Garibay, Efraín Cáceres Cordero, Victoriano Garzón Aguilar, Víctor Donayre Miranda, Carlos García Cristóbal, Diego Silva Inuma, Pedro Del Águila Pezo, Eduardo Huamán Yataco, Jhazmin Pomalaza Acosta, Edison González, Héctor Ángeles and Oliver Pereyra. Finally, we would like to thank Jaime Jahncke by his helpful review and corrections of this manuscript.

References

Agrorural. (2023). *Reporte mensual de conservación en islas y puntas guaneras – Enero 2023*. Lima.

Alfaro-Shigueto, J., Valqui, J., & Mangel, J. C. (2011). Nuevo registro de la nutria marina Lontra felina (Molina, 1782) al norte de su distribución actual. *Ecological Applications, 10*, 87.

Apaza, M., & Romero, L. (2012). Distribución y observaciones sobre la población de la nutria marina Lontra felina (Molina 1782) en el Perú. *Revista Peruana de Biología, 19*, 285–298. https://doi.org/10.15381/rpb.v19i3.1064

Apaza, M., Valqui, J., Mangel, J., et al. (2004). *Estado de Conservación de Lontra felina (Molina, 1782) en la Costa Peruana*. Reporte para la Comisión Permanente del Pacífico Sur.

Barocas, A., Farfan, J., Groenendijk, J., et al. (2022). Disturbance-specific behavioral responses of giant otters exposed to ecotourism and extractive activities. *Animal Conservation, 25*, 15–26. https://doi.org/10.1111/ACV.12713

Biffi, D., & Iannacone, J. (2010). Variabilidad trófica de Lontra feline (Molina 1782) (Carnivora: Mustelidae) en dos poblaciones de Tacna (Perú) entre Agosto y Diciembre de 2006. *Mastozoología Neotropical, 17*, 11–17.

Briones-Salas, M., Peralta-Pérez, M. A., & Arellanes, E. (2013). Análisis temporal de los hábitos alimentarios de la nutria neotropical (Lontra longicaudis) en el rio Zimatán en la costa de Oaxaca, México. *Therya, 4*, 311–326. https://doi.org/10.12933/THERYA-13-138

Cabello, R., & Sánchez, G. (2006). *Parámetros ambientales y residuo de contaminantes orgánicos persistentes (COPs) en el litoral costero de Pisco-Paracas, Perú*. Invierno austral 2005.

Catenazzi, A., & Donnelly, M. A. (2008). Sea lion Otaria flavescens as host of the common vampire bat Desmodus rotundus. *Marine Ecology Progress Series, 360*, 285–289. https://doi.org/10.3354/MEPS07393

Coker, R. E. (1919). Habits and economic relations of the guano birds of Peru. *Proceedings of the United States National Museum, 56*, 449–511. https://doi.org/10.5479/SI.00963801.56-2298.449

CONAF. (2008). *Plan de manejo Monumento Nacional Isla Cachagua*. CONAF.

Cuellar-Martinez, T., Huanca Ochoa, A. d. R., Sánchez, S., et al. (2021). Paralytic shellfish toxins in Peruvian scallops associated with blooms of Alexandrium ostenfeldii (Paulsen) Balech & Tangen in Paracas Bay, Peru. *Marine Pollution Bulletin, 173*, 112988. https://doi.org/10.1016/j.marpolbul.2021.112988

Cursach, J. A., Rau, J. R., Ther, F., et al. (2012). Sinantropía y conservación marina: el caso del chungungo Lontra felina en el sur de Chile. *Revista de Biología Marina y Oceanografía, 47*, 593–597. https://doi.org/10.4067/S0718-19572012000300022

Fajardo, J., Lessmann, J., Bonaccorso, E., et al. (2014). Combined use of systematic conservation planning, species distribution modelling, and connectivity analysis reveals severe conservation gaps in a megadiverse country (Peru). *PLoS One, 9*, e114367. https://doi.org/10.1371/journal.pone.0114367

FAO. (2012). *Estado de las Áreas Marinas y Costeras Protegidas en América Latina*. Enhanced Reader.

Geldmann, J., Barnes, M., Coad, L., et al. (2013). Effectiveness of terrestrial protected areas in reducing habitat loss and population declines. *Biological Conservation, 161*, 230–238. https://doi.org/10.1016/J.BIOCON.2013.02.018

Gray, C. L., Hill, S. L. L., Newbold, T., et al. (2016). Local biodiversity is higher inside than outside terrestrial protected areas worldwide. *Nature Communication, 71*(7), 1–7. https://doi.org/10.1038/ncomms12306

Huayanca, R., Figueroa, C., & Pisconte, J. (2017). *Population variation on shorebirds and human related disturbances in La Aguada and Santo Domingo beaches*. Paracas National Reserve.

INRENA. (1996). *Reserva Nacional de Paracas. Plan Maestro*. INRENA.

INRENA. (1997). *Plan Maestro Lagunas de Mejia*. INRENA.

INRENA. (2002). *Reserva Nacional de Paracas. Plan Maestro 2003–2007*. INRENA.

Mangel, J. C., Whitty, T., Medina-Vogel, G., et al. (2011). Latitudinal variation in diet and patterns of human interaction in the marine otter. *Marine Mammal Science, 27*, E14–E25. https://doi.org/10.1111/J.1748-7692.2010.00414.X

Medina-Vogel, G., Merino, L. O., & Monsalve Alarcón, R. d. J. (2008). Coastal–marine discontinuities, critical patch size and isolation: Implications for marine otter conservation. *Animal Conservation, 11*, 57–64. https://doi.org/10.1111/J.1469-1795.2007.00151.X

Millennium Ecosystem Assessment. (2005). *Ecosystems and human well-being: Synthesis*. Island Press.

Ministerio de Agricultura. (1979). *Reserva Nacional de Paracas. Plan Maestro*. Ministerio de Agricultura.

Ministerio de Agricultura. (2006). DS 043-2006-AG.

Ormeño, M., & Anchante, H. (2009). *Distribución, comportamiento y conservación de la nutria marina (Lontra felina) en la Reserva Nacional de Paracas, Perú*. Technical report presented to INRENA (Instituto Nacional de Recursos Naturales, National Institute of Natural Resources).

Ortiz-Alvarez, C., Alfaro-Shigueto, J., & Mangel, J. C. (2021). Insights into marine otter (Lontra felina) distribution along the Peru coastline. *Marine Mammal Science, 37*, 993–1007. https://doi.org/10.1111/MMS.12789

Pitcher, G. C., Aguirre-Velarde, A., Breitburg, D., et al. (2021). System controls of coastal and open ocean oxygen depletion. *Progress in Oceanography, 197*, 102613.

Pizarro Neyra, J. (2008). Mortality of the marine otter (Lontra felina) in Southern Peru. *IUCN Otter Specialist Group Bulletin, 25*(2), 94–99.

Poblete, A. A., Górski, K., Moscoso, J., et al. (2019). Estimación de dietas del chungungo Lontra felina (Molina, 1782) en dos localidades de la región del Biobío, Chile. *Gayana (Concepción), 83*, 1–9. https://doi.org/10.4067/S0717-65382019000100001

Quispe, D., Graco, M., Correa, D., et al. (2010). Spatio-temporal variability of hydrophysical conditions at Bahia Independencia, Pisco-Perú, from 1995 to 2004. *Ecological Applications, 9*, 9–18.

Republica del Perú. (1997). Ley de Áreas Naturales Protegidas Ley N° 26834. *Diario Oficial El Peruano, 6215*, 150721–157025.

Reyes, J. C. (1992). Informe nacional sobre la situación de los mamíferos marinos en Perú. *Informes y estudios el Programa Mares Regionales del PNUMA,145*, 1–21.

Sanchez-Scaglioni, R. (1990). *Observaciones ecológicas en el gato marino Lutra felina Isla San Gallan, Reserva Nacional de Paracas*. En Centro de Investigaciones sobre Mamíferos Marinos (CIMMA). 4ta Reunión de Trabajo de especialistas en Mamíferos Acuáticos de América del Sur.

Sanchez-Scaglioni, R. (1992). *Distribución y abundancia del gato marino Lutra felina en el Departamento de Ica*. Programa de Subvenciones WCI-APECO, 29pp. Informe inedito.

Sanchez-Scaglioni. (2004). *Evaluación de la distribución y abundancia del gato marino Lontra felina en la Reserva Nacional de Paracas durante el invierno del 2004*. Informe presentado a APECO.

SERNANP. (2016). *Plan Maestro de la Reserva Nacional de Paracas, 2016–2020*. SERNANP.

SERNANP. (2019). *Plan maestro de la Reserva Nacional San Fernando 2019–2024*. http://sis.sernanp.gob.pe/biblioteca/?publicacion=1802. Accessed 25 Mar 2022.

SERNANP. (2020). *Resolución Presidencial N° 104-2020-SERNANP*. Aprueban investigaciones prioritarias de las veinte y cinco (25) Áreas Naturales Protegidas del Sistema Nacional de Áreas Naturales Protegidas por el Estado – SINANPE.

Taylor, M. H., Wolff, M., Mendo, J., & Yamashiro, C. (2008). Changes in trophic flow structure of Independence Bay (Peru) over an ENSO cycle. *Progress in Oceanography, 79*, 336–351. https://doi.org/10.1016/j.pocean.2008.10.006

Valqui, J. (2012). The marine otter Lontra felina (Molina, 1782): A review of its present status and implications for future conservation. *Mammalian Biology, 77*, 75–83.

Valqui, J., Hartl, G. B., & Zachos, F. E. (2010). Non-invasive genetic analysis reveals high levels of mtDNA variability in the endangered South-American marine otter (Lontra felina). *Conservation Genetics, 11*, 2067–2072.

Zavalaga, C. B., & Alfaro-Shigueto, J. (2015). Unveiling an important humboldt penguin (Spheniscus humboldti) breeding colony in Perú and the need for its protection against the potential impact of guano harvest. *38*, 302–307. https://doi.org/10.1675/063.038.0311.

Chapter 6
The Marine Otter in Continental and Andean Habitats

Joaquín A. Ugarte-Núñez and Ulrich Zanabria-Alarcón

Abstract The marine otter, known as chungungo or huallaque (*Lontra felina*), is considered a marine species inhabiting rocky habitats along the Pacific coast from northern Peru to Tierra del Fuego in southern Chile. In this study, we present the continental records of *L. felina* in riverine and lacustrine environments. In the Cotahuasi-Ocoña River basin (Peru), the furthest record from the Pacific coast is found at a higher altitude in the Sipia locality, 171 km from the coast, and at an elevation of 1994 m above sea level. In the Colca-Majes River basin, the furthest record is in the Mamacocha Lagoon in the Caylloma province, 157 km from the sea at an altitude of 1702 m above sea level. Additionally, other direct and indirect records were recorded from the rivers of southwestern Peru. The presence of this species, mainly in two of the largest rivers on the Pacific slope in southwestern Peru, may be due to the extremely canyonized configuration of the Cotahuasi and Colca Rivers. This geographical characteristic could serve as a refuge for this threatened species in addition to providing an abundant food supply. This assertion is discussed based on a review of diet studies, emphasizing the need for conservation actions throughout the continental distribution of marine otters. Some activities, such as dam construction, can severely affect its presence and sustainability in these environments.

Keywords Huallaque · Rivers · Lagoons · Canyons · Threats · Conservation

6.1 Introduction

The marine otter, also known as chungungo or huallaque (*Lontra felina* Molina, 1782), is the smallest species within the genus and the sole marine otter along the Pacific coast of South America (Estes, 1986). It is classified as Endangered (EN) by

J. A. Ugarte-Núñez (✉) · U. Zanabria-Alarcón
Asociación para la Conservación y el Desarrollo Sostenible Sallqa Perú, Arequipa, Peru
e-mail: joaquin.ugarte@sallqaperu.org; ulrich.zanabria@sallqaperu.org

L. Ayala et al. (eds.), *Marine Otter Conservation*,
https://doi.org/10.1007/978-3-031-53931-2_6

the International Union for Conservation of Nature (Valqui & Rheingantz, 2021). Also, Peruvian legislation classified marine otter as threatened species in Peru (MINAGRI, 2014). Its distribution spans the cold temperate region of the Pacific coast from Peru to Chile, with the largest population residing in Chile (Yoxon & Yoxon, 2014). Several references regarding its northern distribution limit in Peru, with the most widely accepted limit being Chimbote in the Ancash Department (9°10′S) (Valqui, 2012). However, Schweigger (1964) reported the species as far north as Isla Lobos de Tierra in the Piura Department (6°27′S). Apaza and Romero (2012), in their review of Punta Aguja (05°47′S) in Piura to Tacna (border with Chile 18°21′S), confirmed the northern limit at Chimbote, casting doubt on the report in Isla Lobos de Tierra. They considered the records in the port of Huanchaco (08°04′S) (Alfaro-Shigueto et al., 2011) and in Puerto Salaverry (8°13′S) (Santillán & Caro, 2007), La Libertad department, as marginal records in the transition zone north of the optimal habitat. These records are the current northern limit, with a possible marginal distribution further north or with occasional colonization events (Valqui, 2012). The southern limit is Tierra del Fuego in southern Chile and Isla de los Estados (54°48′S) in southern Argentina (Larivière, 1998).

L. felina have been recognized as an exclusively marine species with a preference for rocky coastal habitats (Larivière, 1998; Eisenberg & Redford, 1999; Boher, 2005; Córdova et al., 2009; Valqui et al., 2010; Vianna et al., 2010; Apaza & Romero, 2012; Yoxon & Yoxon, 2014; Sanino & Meza, 2016). *L. felina* may be absent from several hundred kilometers of the coast because of its naturally fragmented habitat (Eisenberg & Redford, 1999; Vianna et al., 2010; Valqui, 2012). It is an alternation between very heterogeneous habitats considered suitable (rocky patches with caves, sometimes piers, shipwrecks, or abandoned fishing boats) and unsuitable habitats such as sandy beaches or rocky shores without caves (Eisenberg & Redford, 1999; Vianna et al., 2010; Valqui, 2012). However, Rozzi and Torres-Mura (1990) described the species using sandy beaches, indicating that this species is currently associated with fragments of exposed rocky coastlines. This shift in habitat uses by increasing human occupation, utilization, and exploitation of coastal resources (Rozzi & Torres-Mura, 1990).

Few records have been reported *L. felina* in estuaries and some meters upstream (Estes, 1986). *L. felina* was reported in the Camaná and Ocoña Rivers, feeding on shrimp, up to 40 km from the coast and at 650 m altitude[1] (Hvidberg-Hansen, 1970; Tello, 1972; Brownell, 1978; Viacava et al., 1978). Valqui (2012), citing Apaza et al. (2004), considers that these records no longer occur because of the growth of human population and increased surveillance in shrimp farming, which probably causes the current presence of *L. felina* predominantly in marine areas. Other studies provided confusing reports of otter presence in the Cotahuasi River as *Lontra longicaudis* (Ugarte & Salazar, 1998). Later reported by Zeballos Patrón et al. (2001) in the list of mammals of Arequipa, which was corrected by Medina et al. (2018) and SERNANP (2019) to be *L. felina*. Likewise, there are more recent studies on the diet

[1] It is remarkably that the altitude is 235 m from Google Earth.

of *L. felina* in the river areas of the Ocoña and Colca-Majes River basins (Medina et al., 2018; Ruelas Cabana et al., 2018; Zanabria et al., 2021).

This study reports direct and indirect records as well as anecdotal evidence of marine otters in river habitats at distances and altitudes exceeding those recorded in the literature. The discussion includes studies on the diet of *L. felina* in the Colca-Majes Basin and explores why otters are found in these basins and the pressures they face.

6.2 Materials and Methods

6.2.1 Study Site

The middle and lower Andean regions of the rivers in the Pacific Basin of southwestern Peru primarily originate from the Andean highlands (>5000 m). This area encompasses the two of the deepest canyons in the world (Rademaker, 2014). One corresponds to the Cotahuasi River basin, which later forms the Ocoña River by joining the Marán and Arma Rivers. The second corresponds to the Colca River basin, which forms the Majes River downstream by joining the Capiza River and other minor tributaries. Subsequently, upon merging with the Pucayura River, the Camaná River was formed. These two rivers (Ocoña and Majes-Camaná) flow into the Pacific Ocean in the Arequipa department, with the lower sections having a higher human population of several thousand inhabitants and hundreds in the upper basin sections, i.e., the Cotahuasi and Colca Rivers (INEI, 2018). Agriculture is the primary activity, and shrimp extraction is performed artisanally. However, shrimp extraction contribute to national production nearly 80% (Pinazo et al., 2020). The Cotahuasi Basin is included entirely in the Cotahuasi Sub-basin Landscape Reserve. The rest of the rivers belonging to the Pacific Basin that were reviewed and do not present a significantly canyon-like configuration are the Acarí, Yauca, Quilca, and Tambo Rivers in the Arequipa Department, the Osmore or Moquegua River in the Moquegua Department, and the Locumba, Sama, and Caplina Rivers in the Tacna Department (Fig. 6.1). These rivers belong to the Pacific slope of Peru and are characterized by marked seasonal variability between the summer and winter months. Summer months, considering the flood period, experienced the highest water flows, whereas winter periods had the lowest flows (ANA, 2010; Oliveira et al., 2015). The Ocoña and Camaná-Majes Basins are significant among the 53 rivers of the Pacific slope because they have the largest area and flows that exceed 100 m^3/s during flood periods (Gutierrez Lope et al., 2022). They also exhibit annual variability, with the highest precipitation values from November to April and the lowest values from June to July (ANA, 2015a, 2015b).

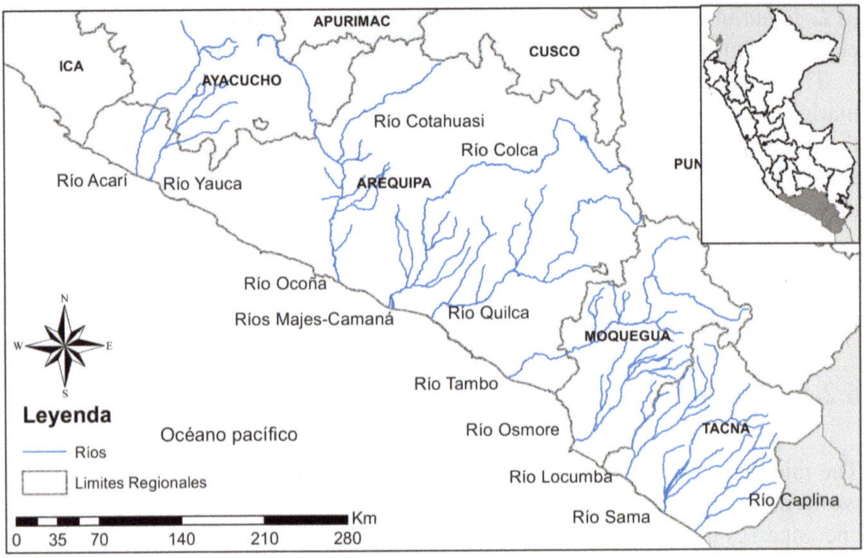

Fig. 6.1 Rivers assessed to determine the continental presence of *Lontra felina*

6.2.2 *Occasional Sightings*

Based on chronicles and occasional sightings between 1997 and 2018, the authors traversed the banks of the Cotahuasi-Ocoña, Colca-Majes, Quilca, and Tambo River basins, covering approximately 350 km, of which 290 km corresponds to the first two basins, to observe and gather further evidence of the presence of *L. felina*. The records included the following: date, location, altitude, number of observed individuals, habitat type (rocky area, river, shore), and activity (feeding, swimming, and resting). River records are those located within the vital domains of freshwater in the coastal region (<1000 m above sea level) and the western slopes of the Andes (above 1000 m above sea level), based on Koepcke and Koepcke's (1968) vital domain classification: riverbeds with current, rocky, or gravelly riverbanks, or low vegetation and riparian forests. To be considered river records in freshwater, estuarine records were excluded because of the saline and mixed conditions with seawater up to 50 m in altitude, characteristics that confer it as the area of greatest shrimp reproduction in this type of water (Pinazo et al., 2020) and as an area of more frequent otter sightings (Estes, 1986).

6.2.3 *Interviews*

Twenty-two residents engaged in activities such as fishing or farming near rivers were interviewed to obtain information about the presence of *L. felina* and their perception of the species. The residents were all farmers, engaged in fishing only for family consumption and not commercially. Non-systematic interviews were conducted in informal conversations to avoid refusal to converse due to a certain aversion toward otters (Ugarte & Salazar, 1998). Information sought from local residents pertained to four aspects, all related to the location of the interview: (1) awareness of the presence of "huallaques" or otters in the river, (2) if they see them at a particular time of year and how many are usually seen, (3) if they think there are fewer now than before (40 years ago), and (4) their opinion on the presence of otters in the river.

6.2.4 Literature Review on Diet

A search for published and unpublished information was conducted using Google Scholar, Web of Science, Scopus, ResearchGate, and university repositories. The following keywords in English and Spanish were used: "*Lontra felina*", "marine otter", "diet" and "feeding".

6.3 Results

6.3.1 Sightings and Interviews

A total of 16 direct and anecdotal records of *Lontra felina* were documented between 1997 and 2018 in three basins of the southern Pacific slope of Peru. Nine were direct observations of individuals swimming in rivers and seven were anecdotal records from interviews (Table 6.1, Fig. 6.2).

In the Cotahuasi River, located in the Andean province of La Unión and the Subbasin Landscape Reserve of Cotahuasi, four records have been made (two direct and two anecdotal), one from Sipia, 171 km from the Pacific coast, and at an altitude of 1994 m. This is the highest and farthest direct record from the coast.

In the Ocoña River, downstream from the Cotahuasi River, three additional records were reported (two direct and one anecdotal); one from Chiguay in the province of Camaná was the lowest in altitude (70 m) and distance from the coast (14 km). In the Majes-Colca Basin, eight records were obtained (five direct and three anecdotal). The observation from Mamacocha Lagoon in the province of Castilla (Fig. 6.3) is a direct record that extends up to 157 km from the sea and reaches an altitude of 1702 m, with its waters flowing into the Colca River (bordering Caylloma province). The third basin with records is the Quilca River, with an anecdotal record of 25 km from the coast and at an altitude of 212 m (Fig. 6.2).

Sightings occurred in the farthest locations from the coast, Sipia and Mamacocha Lagoon, in the middle parts of the Andean basins of the Cotahuasi and Colca Rivers, as well as in other downstream locations in the Cotahuasi-Ocoña Basin: Velinga in La Unión province, Lamapampa-Chillihuay in Condesuyos, and Chiguay in Camaná

Table 6.1 Direct and anecdotal records of *Lontra felina* in riverine environments, southwestern Peru

	Locality	Province	Record type	Record year	Latitude	Longitude	Altitude (masl)
1	Sipia	La Unión	Sighting/pelt	1999	15°14′32″	72°58′02″	1983
2	Chaupo	La Unión	Interview	2018	15°15′04″	72°59′31″	1852
3	Velinga	La Unión	Sighting	1997	15°16′43″	73°01′45″	1705
4	Chaucalla	La Unión	Interview / taxidermy otter	1997	15°35′12″	73°05′49″	889
5	Lamapampa-Chillihuay	Condesuyos	Sighting	1999	15°40′30″	73°03′38″	773
6	La Barrera	Condesuyos	Interview	1999	15°46′54″	73°05′03″	652
7	Chiguay	Camaná	Sighting	2016	16°20′40″	73°08′26″	70
8	Laguna Mamacocha	Castilla	Sighting	2015–2018	15°40′39″	72°14′42″	1702
9	Tingo	Castilla/Caylloma	Sighting	2016–2018	15°42′37″	72°14′08″	1463
10	Carihua	Castilla/Caylloma	Sighting/adult female carcass with fetus	2016–2018	15°43′44″	72°14′41″	1341
11	Cañón Andamayo	Castilla	Sighting	2016–2018	15°51′14″	72°24′53″	946
12	Andamayo	Castilla	Sighting	2014–2018	15°51′18″	72°26′44″	886
13	Ongoro	Castilla	Interview	2014	15°59′17″	72°27′59″	713
14	Aplao	Castilla	Interview	2014	16°04′39″	72°28′54″	609
15	Characta	Camaná	Interview	2014	16°30′35″	72°38′28″	114
16	Quilca	Camaná	Interview	2017	16°37′22″	72°18′14″	212

province. In the Colca-Majes Basin, sightings occurred in the Tingo and Carihua sectors at the boundary of the Castilla and Caylloma provinces, Andamayo canyon, and the town of Andamayo (Castilla). A carcass with a fetus was discovered in the Carihua sector (Fig. 6.4). Interviews were conducted in other locations, including Chaupo and Chaucalla in the Cotahuasi Basin, La Barrera in the Ocoña River, Ongoro, Aplao, and Characta in the Majes River, and at the confluence of the Siguas and Vítor Rivers in Quilca (Fig. 6.2). All interviewees claimed that they were familiar with the huallaque (a giant otter). They also mentioned that there was no specific season for sightings but acknowledged that otters were more abundant when rivers swelled during the rainy season between January and March. Likewise, all affirmed that otters were observed less frequently than before. Finally, except for two interviewees in Ayo (Mamacocha Lagoon), all stated that the otter competed with them for river shrimp, reflecting a negative perception.

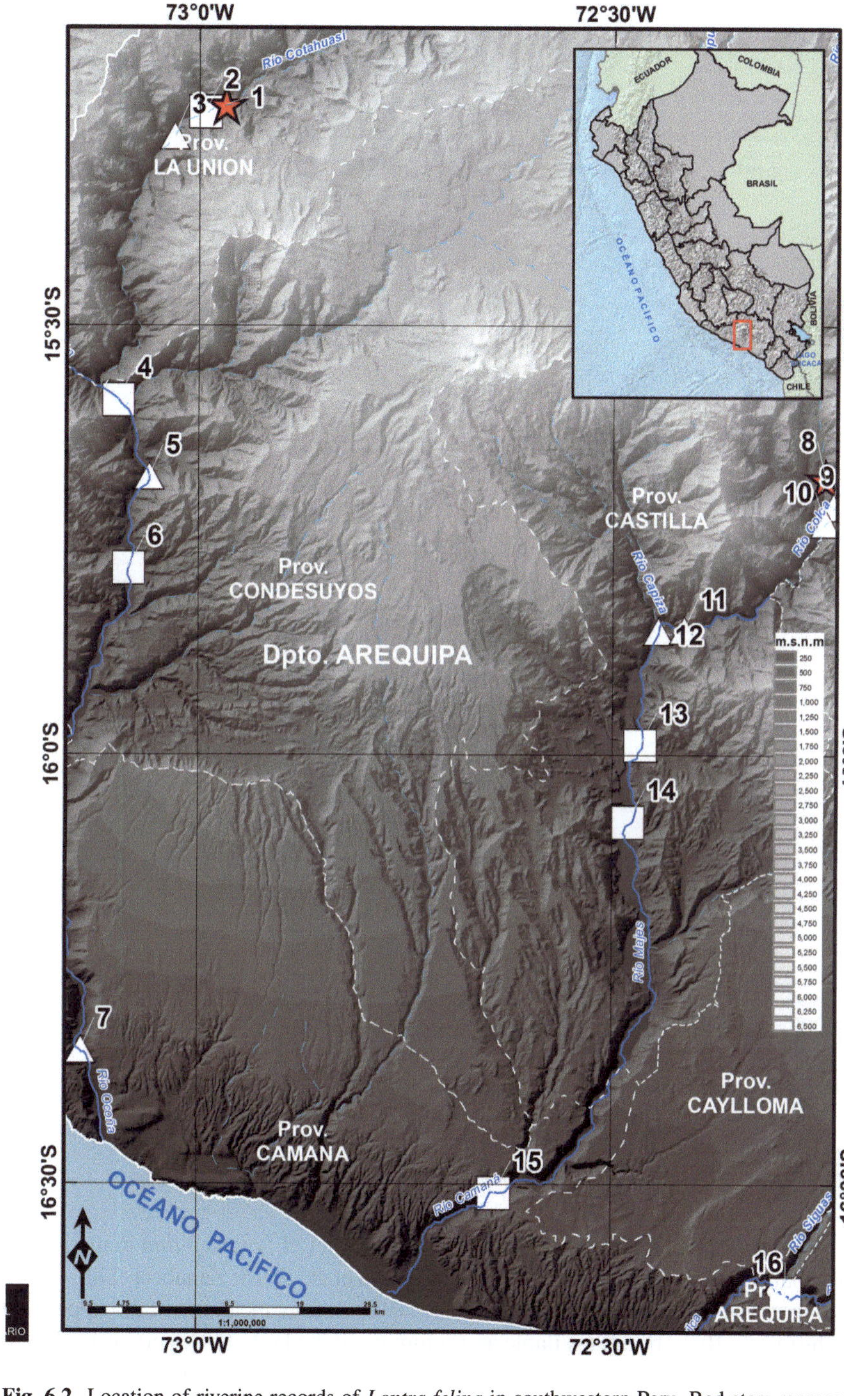

Fig. 6.2 Location of riverine records of *Lontra felina* in southwestern Peru. Red stars represent sightings farthest from the coastline and at higher altitudes, white triangles indicate other sighting records, and white squares denote anecdotal and indirect records (interviews, pelts, and carcasses). Numbers are the locations in Table 6.1

Fig. 6.3 *Lontra felina* individual in Mamacocha Lagoon (January 2015)

Fig. 6.4 Carcass of female *L. felina* with fetus found in the locality of Carihua (Colca River)

6.3.2 Literature Review on Diet

There are two specific unpublished studies on the diet of *Lontra felina* in its riverine distribution (Medina et al., 2018; Ruelas Cabana et al., 2018; Zanabria et al., 2021). The first was conducted by Medina et al. (2018) in four locations in the upper part of the Ocoña River basin, available as a summary of the IV Latin American Congress and VIII Bolivian Congress of Mastozoology (2018) to provide complementary information for the Environmental Impact Assessment of the OCO 2010 Hydroelectric Power Plant (MINEM, 2021). The second study was in Mamacocha

Lagoon and four locations in the Colca-Majes Rivers presented as a poster by Ruelas Cabana et al. (2018) at the same Congress; also, Zanabria et al. (2021) provides more detailed information from the same research in the Workshop "Diagnosis for the proposal of the National Conservation Plan for the Otters of Peru: Marine Otter."

The continental diet has a wide variety of food items that differ between riverine and lentic water habitats (Zanabria et al., 2021). In river habitats, they primarily feed on shrimp (*Cryphiops caementarius*), whereas in the lagoon, prey items are more diverse than in the river (11 versus 6), with Odonata (*Anisoptera*) being the most prominent, including a strong component of *Basilichthys* cf. *archaeus* (river silverside), mollusks of the genus Physella, and small vertebrates such as rodents of the genus *Phyllotis*.

6.4 Discussion

This study reports the presence of *Lontra felina* in three river basins of southern Peru. These records significantly expand the previously reported (Hvidberg-Hansen, 1970; Tello, 1972; Viacava et al., 1978; Brownell, 1978). All of them considered historical records that are no longer present (Valqui, 2012). Specially, the records of the Cotahuasi River (Sipia) and Mamacocha in the Colca Basin, both over 100 km from any previous historical distribution report and at an altitude of over 1500 m above sea level. Likewise, these records confirm and expand upon the information and locations of *L. felina* in the Cotahuasi River, previously reported as *L. longicaudis* (Ugarte & Salazar, 1998), and the diet studies by Medina et al. (2018) in the Ocoña River and Zanabria et al. (2021) for the Colca-Majes Basin.

L. felina is the only aquatic mammal in these areas, historically identified as "huallaque" in its riverine distribution in the Cotahuasi and Marán River basins (V. Pacheco, pers. com 1998). Furthermore, this continental name would be a regionalism only found in Arequipa, but with an origin on the coast, which could indicate that the name has been maintained in the Andes and may have changed on the coast (Pizarro-Neyra, 2022). Residents interviewed near the sea (Chiguay and Characta) indicate that they know it as "chungungo," and further upstream, in the rivers, they call it "huallaque," even suggesting it might be a different species.

Most records are in the Cotahuasi-Ocoña and Colca-Majes river basins, where the marine otter is known as "huallaque," showing that its distribution and presence have been widely known by the residents of these basins. However, interviewees indicated that in recent decades, the number of observed individuals has decreased, as in the 1950s, dozens of them used to ascend the rivers. Even considering the interviewees' perception, which could overestimate the reported number of individuals, it was concluded that the species was sighted more frequently and in greater quantities in the past.

There are no rivers southwest of Peru to the north or south of the two previously described basins where the species has been recorded, in addition to the record in the Quilca River, 25 km from the coast. In the Tambo River (between Moquegua and Arequipa), which has a considerable and permanent flow, there are no records, except in the estuary (obs. pers. Joaquin Ugarte). The highest and farthest record in the town of Sipia in the Cotahuasi Basin may have a waterfall of the same name as a natural barrier, as the town is located at the foot of this waterfall (approximately 150 m).

The otter reproduction in its Andean continental distribution is corroborated by the discovery of a carcass in the town of Carihua (1341 m above sea level) with a fetus inside. Additionally, residents interviewed in Velinga (Cotahuasi River) claimed to have seen offspring. They indicate that the species reproduces in that area (1600 m above sea level), meaning that it also reproduces in its riverine distribution area and does not only enter rivers opportunistically to feed, because of the abundance of shrimp (Hernández, 1960, cited by Pizarro-Neyra, 2022).

While this study did not focus on the diet of *Lontra felina*, the sighting locations in the Ocoña Basin (Medina et al., 2018) and Colca-Majes served for the diet analysis developed by Zanabria et al. (2021), providing important information for future conservation actions for the species. The differences found between the riverine habitat (Ocoña and Colca-Majes Rivers, Fig. 6.5) and lacustrine habitat (Mamacocha Lagoon, Fig. 6.6) could indicate that its trophic niche may have a greater breadth in the lagoon due to opportunism and the availability of prey (absence of abundant river shrimp) according to Zanabria et al. (2021). This behavior is observed in its marine distribution, where there are references to feeding on birds and small mammals (Larivière, 1998) as well as records in *Lontra longicaudis* of feeding on birds and herpetofauna (Pardini, 1998). Residents of the Cotahuasi-Ocoña and Majes Basins perceive *L. felina* as a competitor for shrimp (*Cryphiops caementarius*), a species that is highly appreciated and consumed in the coastal rivers of southern Peru. The rivers with a higher shrimp population in Peru are Majes and Ocoña, along with Tambo to the south, because these rivers carry a higher water flow (Pinazo et al., 2020). This negative perception of residents near the river and shrimp consumers, combined with historical hunting for the pelt of rugs (Chaucalla and Sipia) and other ornamental objects (Ugarte & Salazar, 1998), poses a risk to the survival of this species in these areas. However, in areas near the Mamacocha Lagoon (Ayo), residents do not have this negative perception, probably because of the absence of shrimp in the lagoon (Zanabria et al., 2021).

This work validates the presence of *L. felina* in the Sub-basin Landscape Reserve of Cotahuasi. Although recognizing it as part of the faunal richness in its Master Plans, does not prioritize it since its creation in 2005, despite the protected natural area's goal of harmonizing conservation with the sustainable development of the La Unión province. According to its known distribution in Peru, this species is also found in marine protected natural areas (National Reserve System of Islands, Islets and Guano Points, Paracas National Reserve, and San Fernando National Reserve) (Valqui & Rheingantz, 2021; Pizarro-Neyra, 2023). Thus, current conservation measures are focused on addressing threats in its marine distribution rather than riverine. Although monitoring *L. felina* has recently been included in the Sub-basin

Fig. 6.5 Habitat of *Lontra felina* in the Colca River. Note the strip of vegetation where the burrows and resting areas of the otters are located. (Photo: Gonzalo Medina-Vogel)

Landscape Reserve of Cotahuasi (SERNANP, 2023), it is necessary to include management actions for this species in its Master Plan to contribute to its conservation.

Finally, it is highly likely that the presence of *L. felina* in the Cotahuasi and Colca River basins is due to their extremely rugged configuration and the low human population in these sections of the basins, providing some protection to the species. Thus, these basins could constitute the last or only refuge for riverine distribution, considering the importance of the abundance of food resources such as river shrimp and possibly fish, such as trout and silversides (Ugarte & Salazar, 1998). These factors, refuge and a variety of food availability, would indicate that the marine otter is an opportunistic species that selects its prey based on availability rather than energy contribution (Bastida et al., 2007; Zanabria et al., 2021), and factors such as proximity to its prey and the degree of exposure of the feeding site could be significant factors determining its distribution (Medina-Vogel et al., 2007; Villegas et al., 2007).

Any intervention that disrupts the riverine distribution, the probable route of genetic exchange, and the establishment of this critically endangered and charismatic species, such as the construction of infrastructure that dams water, could have a substantial impact on these populations. Their ecology and freshwater biology are still poorly understood. Currently, there are two hydroelectric projects with approved environmental management instruments. The first is the OCO 2010 Hydroelectric Power Plant on the southern border of the Sub-basin Landscape Reserve of

Fig. 6.6 Habitat of *Lontra felina* in the Mamacocha Lagoon. (Photo: Gonzalo Medina-Vogel)

Cotahuasi (MINEM, 2021), where SERNANP is currently monitoring *L. felina*. The second is the Laguna Azul Hydroelectric Power Plant at the overflow of Mamacocha Lagoon in the Colca Basin (ARMA, 2017). Therefore, it is recommended to conduct ecological and genetic assessments of its riverine populations, vertical movements, conduct further studies on feeding and reproduction, and take measures to mitigate identified threats. This includes raising awareness among residents, their authorities, state entities, and ensuring the proper management of infrastructure in these areas.

Acknowledgement This chapter is built upon the article: La nutria marina (Lontra felina) en los Andes del sur del Perú. Joaquín Ugarte-Núñez. Ciencia y Desarrollo 20(1), 71–77 https://revistas.unjbg.edu.pe/index.php/cyd/article/view/1109. Licensed under CC 4.0 http://creativecommons.org/licenses/by/4.0/.

References

Alfaro-Shigueto, J., Valqui, J., & Mangel, J. C. (2011). Nuevo registro de la nutria marina Lontra felina (Molina, 1782) al norte de su distribución actual. *Ecological Applications, 10*, 87.

ANA. (2010). *Estudio de máximas avenidas en las cuencas de la zona centro de la Vertiente del Pacífico: Informe final*. ANA.

ANA. (2015a). *Evaluación de recursos hídricos en la cuenca Camaná – Majes – Colca*. Autoridad Nacional del Agua.

ANA. (2015b). *Evaluación de recursos hídricos en la cuenca de Ocoña*. Autoridad Nacional del Agua.

Apaza, M., & Romero, L. (2012). Distribución y observaciones sobre la población de la nutria marina Lontra felina (Molina 1782) en el Perú. *Revista Peruana de Biología, 19*, 285–298. https://doi.org/10.15381/rpb.v19i3.1064

Apaza, M., Valqui, J., Mangel, J., et al. (2004). *Estado de Conservación de Lontra felina (Molina, 1782) en la Costa Peruana*. In: Reporte para la Comisión Permanente del Pacífico Sur, Lima, 23 p.

ARMA. (2017). Informe Técnico Sustentatorio (ITS) para la Modificación de componentes del Proyecto Central Hidroeléctrica Laguna Azul. Resolución Sub Gerencial Regional N. 0 163–2017- GRA/ARMA-SGCA. Autoridad Regional Ambiental (ARMA) - Gobierno Regional Arequipa.

Bastida, R., Rodríguez, D., & Secchi, E. (2007). *Mamiferos Acuaticos de Sudamerica y Antartida*. Vàzquez Mazzini Editores.

Boher, F. (2005). *Ambito de hogar y territorialidad del Chungungo Lontra felina (Molina 1782) en el litoral central de Chile, Quintay*. Universidad Austral de Chile.

Brownell, J. R. L. (1978). *Ecology and conservation of the marine otter, Lutra felina*. In: Duplaix, N. (Ed.), Otters: Proceedings of the first working meeting of the otter specialist group: 104–106. International Union for Conservation of Nature and Natural Resources, Morges, Switzerland.

Córdova, O., Rau, J. R., Suazo, C. G., & Arriagada, A. (2009). Estudio comparativo de la ecología alimentaria del depredador de alto nivel trófico Lontra felina (Molina, 1782) (Carnivora: Mustelidae) en Chile. *Revista de Biología Marina y Oceanografía, 44*, 429–438. https://doi.org/10.4067/s0718-19572009000200016

Eisenberg, J. F., & Redford, K. H. (1999). *Mammals of the neotropical: The central neotropics*. The University of Chigago Press.

Estes, J. A. (1986). Marine otters and their environment. *Ambio, 15*, 181–183.

Gutierrez Lope, L., Lavado-Casimiro, W., & Felipe-Obando, O. (2022). *Estimación de hidrogramas horarios de caudales en la vertiente del pacífico, estudio final*. Servicio Nacional de Meteorología e Hidrología del Perú.

Hvidberg-Hansen, H. (1970). *A survey of the coast otter (Lutra felina) Molina in the Peruvian rivers of Camana and Ocoña*. Informe Técnico, FAO–UNDP/SP, Lima, Perú.

INEI. (2018). *Arequipa: Resultados definitivos. Tomo I* (pp. 425–430). Caracteristicas de la población.

Koepcke, H., & Koepcke, M. (1968). División ecológica de la Costa peruana. Facultad de Pesquería de la Universidad Agraria, La Molina. Boletín de Extensión No11: 3–22. Lima

Larivière, S. (1998). Lontra felina. *Mammalian Species, 575*, 1. https://doi.org/10.2307/3504377

Medina, C., Díaz, D., Málaga, B., et al. (2018). Dieta de Lontra felina (Carnívora, Mustelidae) en la parte alta del río Ocoña, Arequipa–Perú. In *IV Congreso Latinoamericano, VIII Congreso Boliviano de Mastozoología; 10–13 Jul 2018*. Colección Boliviana de Fauna (Museo Nacional de Historia Natural-Instituto de Ecología Universidad Mayor de San Andrés).

Medina-Vogel, G., Boher, F., Flores, G., et al. (2007). Spacing behavior of marine otters (Lontra felina) in relation to land refuges and fishery waste in Central Chile. *Journal of Mammalogy, 88*, 487–494. https://doi.org/10.1644/06-MAMM-A081R1.1

MINAGRI. (2014). *Decreto Supremo que aprueba la actualización de la lista de clasificación y categorización de las especies amenazadas de fauna silvestre legalmente protegidas*. MINAGRI.

MINEM. (2021). *Resolución Directoral N.° 0145-2021-MINEM-DGAAE – Normas y documentos legales – Ministerio de Energía y Minas*. MINEM.

Oliveira, N., Henriques, A., Miodonski, J., et al. (2015). Seabird bycatch in Portuguese mainland coastal fisheries: An assessment through on-board observations and fishermen interviews. *Global Ecology and Conservation, 3*, 51–61. https://doi.org/10.1016/j.gecco.2014.11.006

Pardini, R. (1998). Feeding ecology of the neotropical river otter Lontra longicaudis in an Atlantic Forest stream, South-Eastern Brazil. *Journal of Zoology, 245*, 385–391. https://doi.org/10.1111/j.1469-7998.1998.tb00113.x

Pinazo, K., Incio Pérez, A., & Campos León, S. (2020). *Pesquería y condición biológica del camarón Cryphiops caementarius (Molina, 1782) en ríos de Arequipa-Perú, 2016–2018*. Instituto del Mar del Perú.

Pizarro-Neyra, J. (2022). ¿Ansumo, huallaque o chungungo?, precisiones sobre los nombres comunes de la nutria marina en el Perú. *Bol la Acad Peru la Leng, 71*, 465–476. https://doi.org/10.46744/bapl.202201.015

Pizarro-Neyra, J. (2023). The marine otter Lontra felina (Molina, 1782) (Mammalia: Carnivora: Mustelidae) along the marine protected areas in Peru. *Journal of Threatened Taxa, 15*, 22510–22514. https://doi.org/10.11609/jott.7566.15.1.22510-22514

Rademaker, K. (2014). Late Ice-Age human settlement of the high-altitude Peruvian Andes. *Mitteilungen der Gesellschaft für Urgeschichte, 23*, 13–36.

Rozzi, R., & Torres-Mura, J. C. (1990). Observaciones del chungungo (Lutra felina) al sur de la isla Grande de Chiloé: antecedentes para su conservación. *Medio Ambient, 11*, 24–28.

Ruelas Cabana, C., Guillén, F., Huamani, S., et al. (2018). Estudio de la dieta de la nutria marina "Lontra felina" (Molina 1782) en la cuenca de los ríos Colca-Majes y en la laguna Mamacocha, Arequipa, Perú. In *IV Congreso Latinoamericano y VIII Boliviano de Mastozoología. Bolivia.*

Sanino, G. P., & Meza, M. I. (2016). Ecología Trófica y Simpatría de Nutrias (Lontra felina y Lontra provocax) en la Reserva Añihué, Patagonia Chilena. *Boletín Museo Nacional de Historia Natural, 65*, 279–289. https://doi.org/10.54830/bmnhn.v65.2016.98

Santillán, L., & Caro, K. (2007). Mamíferos Marinos en el Puerto de Salaverry y Zonas Adyacentes. In *1er Congreso Nacional de Ciencias del Mar del Perú.*

Schweigger, E. (1964). *El litoral Peruano* (2nd ed.). Gráfica Morsom S.A.

SERNANP. (2019). *Plan Maestro de la Reserva Paisajística Subcuenca del Cotahuasi, periodo 2019–2023*. Resolución Presidencial N° 079-2019-SERNANP.

SERNANP. (2023). Resolución Directoral N.° 004/2023-SERNANP-DGANP: Aprobar el "Protocolo de Monitoreo de la abundancia para la nutria (Lontra felina) en la Reserva Paisajística Subcuenca del Cotahuasi".

Tello, E. (1972). *Anotaciones sobre el Camarón*. Documenta, Ministerio de Pesquería N°18; 5–9.

Ugarte, J., & Salazar, E. (1998). Estudio de la biodiversidad vegetal y animal Cuenca del Cotahuasi Riqueza faunística Zona II. In *Asociación Especializada para el Desarrollo (AEDES).*

Valqui, J. (2012). The marine otter Lontra felina (Molina, 1782): A review of its present status and implications for future conservation. *Mammalian Biology, 77*, 75–83.

Valqui, J., & Rheingantz, M. L. (2021). Lontra felina, marine otter. In *IUCN red list threat. Species 2021*. https://www.iucnredlist.org/species/12303/95970132. Accessed 21 Dec 2023.

Valqui, J., Hartl, G. B., & Zachos, F. E. (2010). Non-invasive genetic analysis reveals high levels of mtDNA variability in the endangered South-American marine otter (Lontra felina). *Conservation Genetics, 11*, 2067–2072. https://doi.org/10.1007/s10592-010-0069-9

Viacava, M., Aitken, S. R., & Llanos, J. (1978). *Estudio del Camaron en el Perú 1975–1976*. IMARPE.

Vianna, J. A., Ayerdi, P., Medina-Vogel, G., et al. (2010). Phylogeography of the marine otter (Lontra felina): Historical and contemporary factors determining its distribution. *The Journal of Heredity, 101*, 676–689. https://doi.org/10.1093/JHERED/ESQ088

Villegas, M. J., Aron, A., & Ebensperger, L. A. (2007). The influence of wave exposure on the foraging activity of marine otter, Lontra felina (Molina, 1782) (Carnivora: Mustelidae) in northern Chile. *Journal of Ethology, 25*, 281–286. https://doi.org/10.1007/s10164-006-0024-x

Yoxon, P., & Yoxon, G. (2014). *Otters of the world*. Whittles Publishing.

Zanabria, U., Valqui, J., & Zeballos, H. (2021). Biología de la nutria marina en el sur del Perú: hacia su conservación. In *Taller "Diagnóstico para la propuesta del Plan Nacional de Conservación de las Nutrias del Perú: Nutria Marina"*. SERFOR.

Zeballos Patrón, H., Pacheco, V., & Baraybar, L. (2001). Diversidad y conservación de los mamíferos de Arequipa, Perú. *Revista Peruana de Biología, 8*, 094–104. https://doi.org/10.15381/rpb.v8i2.6564

Chapter 7
Legal Framework and Its Application for the Conservation of the Marine Otter in Chile and Peru

Ricardo G. Correa, Johanna Garay-Rodríguez, and Raúl Sánchez-Scaglioni

Abstract This chapter describes the international agreements and the Chilean and Peruvian national rules and regulations developed to conserve the marine otter, *Lontra felina*. The change and greater precision in the regulations over time is shown, as the conservation status of the species and the increasing risk of its extinction became known. Relevant legal and administrative precedents are also presented in defense of this marine mammal, affected by anthropogenic pressures. The rulings and judicial resolutions are shown and discussed, and attention is drawn to the need to reformulate some regulations and work jointly with the authorities of the Judiciary and public institutions responsible or competent for the conservation of wildlife. We conclude that these actions set a precedent in the conservation and preservation of the marine otter through the tools of law, which are collected for knowledge and experience of future actions.

Keywords *Lontra felina* · Marine otter · Biodiversity loss · Conservation · Preservation · Environmental justice · Environmental law · Sustainability

R. G. Correa
IUCN Species Survival Commission, Otter Specialist Group, Valparaíso, Chile
e-mail: rcorread@gmail.com

J. Garay-Rodríguez (✉)
Independent Researcher, Lima, Peru
e-mail: johagr99gaia@yahoo.com.mx

R. Sánchez-Scaglioni
Asociación Peruana para la Conservación de la Naturaleza, Lima, Peru
e-mail: rsanchezscaglioni@apeco.org.pe

L. Ayala et al. (eds.), *Marine Otter Conservation*,
https://doi.org/10.1007/978-3-031-53931-2_7

7.1 Introduction

The marine otter (*Lontra felina*) (Molina, 1782) is the smallest otter in its genus (Jefferson et al., 2015) and also the smallest marine mammal in the world (Valqui, 2012). It is considered the most recently adapted mammal to the marine environment (Ostfeld et al., 1989). Otters are notable for being slender and playful animals; however, this has not prevented them from being hunted (Foster-Turley et al., 1990).

Globally, hundreds of thousands of otters from different species were captured for their fur (Kruuk, 2006), and marine otters were no exception, as they were hunted and killed until the mid-twentieth century. Thousands of marine otters were captured in Chilean territory, both for local use and for the international market, without any control or regulation (Iriarte & Jaksic, 1986). According to official reports, between 1910 and 1954, 38,000 otter skins from two species, *Lontra felina* and *Lontra provocax*, were exported from Chile (Iriarte & Jaksic, 1986). The capture after 1954 is believed to have stopped due to overexploitation (Castilla, 1981). In the case of Peru, no reports have been found on this matter.

The spatial occupancy range of the marine otter is the narrow coastal strip of Peru, Chile, and part of Argentina (Larivière, 1998). Much of this area is subject to intense human use and interventions (Barragán & Lazo, 2018; Aguilera et al., 2019). While hunting of these species has ceased, other threats now endanger its existence: habitat fragmentation, modification and destruction, attacks by domestic or feral dogs, extraction and fishing of food resources (crustaceans, mollusks, fish, and algae), incidental captures in fishing gear (nets, traps, hooks), disease transmission, among others (Medina-Vogel et al., 2007; Medina-Vogel, 2010; Cursach et al., 2012; Mangel et al., 2010). However, it is important to note that marine otters are synanthropic species, meaning they can inhabit anthropized ecosystems, adapting to environmental conditions created or modified as a result of human activity (Alfaro-Shigueto et al., 2011; Cursach et al., 2012; Mangel et al., 2010).

To ensure the conservation of this and other wild species, various national and international legal instruments have been implemented. Nevertheless, it is urgent to implement additional strategies that minimize or reduce the threats faced by the marine otter. Legal and administrative actions are the defense mechanisms established in legal frameworks to address these threats and achieve species conservation.

This work will detail the most relevant legal instruments and frameworks, as well as changes in legislation in favor of this species over time. Judicial and administrative actions related to events in which marine otters died will also be presented and discussed, aiming to illustrate and understand the civil, administrative, and criminal aspects involved in such cases.

7.2 International and National Legal Framework for the Conservation of the Marine Otter

7.2.1 International Legal Framework

Globally, the International Union for Conservation of Nature (IUCN) is the leading organization for wildlife protection. Two international conventions also stand out in this regard: the Convention on International Trade in Endangered Species of Wild Fauna and Flora (CITES) and the Convention on the Conservation of Migratory Species of Wild Animals (CMS).

The IUCN,[1] an international association of government and non-governmental members, aims to influence, encourage, and assist societies worldwide in conserving the integrity and diversity of nature. It also seeks to ensure that any use of natural resources is done equitably and ecologically sustainably (IUCN, 2022). In 1963, it established the Red List of Threatened Species (Godfrey et al., 2008), and since 1966, it has been regularly published and made available to the general public. The list is continually updated and is the most comprehensive source of information globally regarding the conservation status of animal, fungal, and plant species.

The Red List of Threatened Species has established different categories based on the conservation status of the species, which have been refined over time. Currently, the categories are as follows (IUCN, 2012):

Extinct (EX): A taxon is Extinct (EX) when there is no reasonable doubt that the last individual has died.

Extinct in the Wild (EW): A taxon is Extinct in the Wild (EW) when it is known only to survive in cultivation, in captivity or as a naturalized population (or populations) well outside the past range.

Critically Endangered (CR): A taxon is Critically Endangered (CR) when it is considered to be facing an extremely high risk of extinction in the wild.

Endangered (EN): A taxon is Endangered (EN) when the best available evidence indicates that it is considered to be facing a very high risk of extinction in the wild.

Vulnerable (VU): A taxon is Vulnerable (VU) when the best available evidence indicates that it is considered to be facing a high risk of extinction in the wild.

Near Threatened (NT): A taxon is Near Threatened (NT) when it not qualify for Critically Endangered, Endangered or Vulnerable now, but is close to qualifying for or is likely to qualify for a threatened category in the near future.

Least Concern (LC): A taxon is Least Concern (LC) when it has been evaluated against the Red List criteria and does not qualify for Critically Endangered, Endangered, Vulnerable or Near Threatened.

[1] Originally International Union for Conservation of Nature and Natural Resources.

Data Deficient (DD): A taxon is Data Deficient (DD) when there is inadequate information to make a direct, or indirect, assessment of its risk of extinction based on its distribution and/or population status.

Not Evaluated (NE): A taxon is Not Evaluated (NE) when it has not yet been evaluated against the criteria.

In the IUCN Red List, the marine otter, also known as chungungo or sea cat, is categorized as Endangered (EN) on a global scale according to the latest assessment as of May 5, 2022 (Mangel et al., 2022). The initial inclusion in this international list is believed to have occurred in 1970, also under the Endangered category, although the definition at that time differed from the current one, as noted by Brownell (1978), citing the data from the corresponding Red Book entry for this species:

> *IUCN, 1970. Marine otter: Chingungo Lutra felina. Code No MA/99/Lutra fel, Red Data Book, Survival Service Comission, International Union for Conservation of Nature and Natural Resources.*

The species is not included in the first list published in 1964 (IUCN, 1964) or the one from 1968 (Joslin & Mayanka, 1968). The current justification for considering the marine otter in the Endangered category is as follows (Mangel et al., 2022):

- Its distribution along the Pacific coast is now irregular from Peru to Tierra del Fuego.
- The original distribution area and the global population of the marine otter have decreased due to intensive fur hunting, and populations were nearly exterminated around the southern limit of its previous distribution area, in the regions of Cape Horn and southern Tierra del Fuego, as well as in the northern extremes.
- Most of the species' distribution occurs along the land/sea interface, areas with higher concentrations of human activity in Peru and Chile.
- Its distribution is increasingly restricted due to the growing anthropogenic influence along the coasts.
- The main irreversible factors continually affecting the marine otter population include pollution, accelerated destruction and degradation of habitat, competition for prey, incidental capture, illegal fishing techniques, and mining in coastal habitats.
- These combined factors are resulting in an increase in local extinctions across the species' distribution area.
- The interaction between the marine otter and domestic and invasive species is continuously increasing, with a growing number of cases of pathogen spread to *L. felina*, suspected to impact population numbers.
- There is a continuous reduction in the quality of the *Lontra felina* habitat, and it is suspected that the proportional reduction in the population number will result in a halving of the population between 2006 and 2034.

Eight otter species are listed on the Red List in the categories of Critically Endangered, Endangered, or Vulnerable. In addition to the marine otter, these include the giant otter (*Pteronura brasiliensis*), the Southern river otter (*Lontra provocax*), the sea otter (*Enhydra lutris*), the Asian small-clawed otter (*Aonyx cinereus*), the smooth-coated otter (*Lutrogale perspicillata*), and the hairy-nosed otter (*Lutra sumatrana*).

CITES is an international agreement negotiated among governments with the purpose of ensuring that the international trade of specimens of wild animals and plants does not pose a threat to the survival of species.

It was agreed that Parties (States for which this Convention has entered into force) will not allow trade in specimens of species listed in Appendices I, II, and III of this agreement, with exceptions outlined in the Convention's provisions. The Convention was signed in Washington, D.C., on March 3, 1973, and entered into force on July 1, 1975.[2]

The urgency and importance of signing this agreement became evident following the discovery and disclosure, just days before the Convention meeting, of a vast international network engaged in smuggling and trading skins of threatened and endangered wild animals at the New York airport (Kaplan, 1973). Only one company, Vesely-Forte based in this city, purchased and received 102,000 skins of wild animals, including 15,470 from various otter species and 217 from giant otters (*Pteronura brasiliensis*), largely originating from South America. This underscored the need not only for the signing but also for the ratification of the agreement for prompt entry into force (Fitter, 1973; Gillete, 1973; Kaplan, 1973; Oryx, 1973).

Chile signed the convention on September 16, 1974, and in 1975, the government approved it through Decree Law No. 873 (Decreto Ley N° 873)[3] and Supreme Decree No. 141 of the Ministry of Foreign Affairs (Decreto Supremo N°141 Ministerio de Relaciones Exteriores)[4] ratifying it on February 14, 1975.[5] Thanks to Chile's ratification, the Convention came into effect.

Peru signed the convention on December 30, 1974, approved it on January 21, 1975, through Decree Law No. 21080 (Decreto Ley N° 21,080).[6] On June 27, 1975, Peru ratified the agreement,[7] and it came into effect on September 25, 1975.

---

[2] https://www.eda.admin.ch/eda/fr/dfae/politique-exterieure/droit-international-public/traites-internationaux/depositaire/esp%C3%A8ces-menacees-(cites)/convention-sur-le-commerce-international-des-esp%C3%A8ces-de-faune-et-de-flore-sauvages-menacees-d-extinction.html.

[3] https://www.diariooficial.interior.gob.cl/versiones-anteriores/.
Diario Oficial de la República de Chile, January 28, 1975.

[4] Diario Oficial de la República de Chile, March 25, 1975.

[5] EIDGENOSSISCHES POLITISCHES DEPARTEMENT Notifications aux Etats signataires et adhérents de la CONVENTION SUR LE COMMERCE INTERNATIONAL DES ESPECES DE FAUNE ET DE FLORE SAUVAGES MENACEES D'EXTINCTION, ratification de Equateur, le 11 février, 1975; Chili le 14 février 1975; Uruguay, le 2 avril 1975. Berne, le 9 avril 1975 (Notification, April 9, 1975).

[6] Sistema peruano de información jurídica https://spij.minjus.gob.pe/spij-ext-web/#/detallenorma/H715341, Published in Diario Oficial El Peruano, January 22, 1975.

[7] EIDGENOSSISCHES POLITISCHES DEPARTEMENT Notifications aux Etats signataires ou contractants de la CONVENTION SUR LE COMMERCE INTERNATIONAL DES ESPECES DE FAUNE ET DE FLORE SAUVAGES MENACEES D'EXTINCTION, RATIFICATIONS DU PEROU, DE COSTA RICA ET DE L'AFRIQUE SU SUD, Berne le 23 juillet 1975 (Notification July 23, 1975).

From the outset, the marine otter was listed in Appendix I, which includes all species in danger of extinction that are or may be affected by trade. Trade in specimens of these species must be subject to particularly strict regulation to avoid further endangering their survival and authorized only under exceptional circumstances.

However, during the Second Conference of the Parties held in Costa Rica from March 19 to 30, 1979, Chile proposed an amendment to transfer *Lutra felina* and *Lutra provocax* from Appendix I to Appendix II.

According to the Convention, Appendix II includes:

(a) All species that, while not currently necessarily threatened with extinction, may become so unless trade in specimens of such species is subject to strict regulation to prevent use incompatible with their survival; and,
(b) Those species not affected by trade, which should also be subject to regulation to allow effective control of trade in the species referred to in the preceding subparagraph.

Chile argued that the population of these species had increased, especially in the southern part of the country, estimating between 3500 and 5000 specimens. It reported that, due to the region's complex geographical conditions, control was difficult, and legal protection alone would not guarantee the species' perpetuation. Chile also stated that protection could be enhanced in the future if the species were valued as a potential economic resource.[8] However, the request was rejected.

All otter species are listed in the appendices of CITES. In Appendix I, you will find:

Aonyx capensis microdon (only populations from Cameroon and Nigeria; all other populations are included in Appendix II)
Aonyx cinereus
Enhydra lutris nereis
Lontra felina
Lontra longicaudis
Lontra provocax
Lutra lutra
Lutra nippon
Lutrogale perspicillata
Pteronura brasiliensis

The Convention on the Conservation of Migratory Species of Wild Animals (CMS) or the Bonn Convention brought together the states through which migratory animals pass, as well as the states within the distribution area of these animals. It establishes the legal framework for internationally coordinated conservation measures across the entire migratory range. The Convention was signed in the city of Bonn on June 23, 1979, and entered into force on November 1, 1983.

[8] https://www.speciesplus.net/api/v1/documents/829.

Chile was the first country in the Americas to become a Party to the Convention. It acceded to the Convention on September 15, 1983, and was enacted on October 14, 1981. This act was finally published on December 12, 1981, as Decree 868 MINISTRY OF FOREIGN AFFAIRS (Decreto N° 868 PROMULGA EL CONVENIO SOBRE LA CONSERVACION DE ESPECIES MIGRATORIAS DE LA FAUNA SALVAJE MINISTERIO DE RELACIONES EXTERIORES.[9]

Peru acceded to the Convention on January 24, 1997. This act was officially published on January 28, 1997, as SUPREME DECREE No. 002-97-RE, which stipulates that the Peruvian state adheres to the "Convention on the Conservation of Migratory Species of Wild Animals." (DECRETO SUPREMO N° 002-97-RE Disponen que el Estado peruano se adhiera a la "Convención sobre la Conservación de las Especies Migratorias de Animales Silvestres".)[10]

In CMS, a migratory species is defined as the entire population or any geographically separate part of the population of any species or lower taxon of wild animals, a significant proportion of whose members cyclically and predictably cross one or more national jurisdictional boundaries.[11]

In this Convention "The Parties acknowledge the importance of migratory species being conserved and of Range States agreeing to take action to this end whenever possible and appropriate, paying special attention to migratory species the conservation status of which is unfavourable, and taking individually or in co-operation appropriate and necessary steps to conserve such species and their habitat."[12]

The Convention has two appendices, I and II. Regarding Appendix I, it is stipulated that the Parties "shall endeavour to provide immediate protection for migratory species included in Appendix I".[13]

In Appendix I, migratory species in danger are listed. Specifically, CMS establishes that[14]:

> Parties that are Range States of a migratory species listed in Appendix I shall endeavour:
>
> (a) to conserve and, where feasible and appropriate, restore those habitats of the species which are of importance in removing the species from danger of extinction;
> (b) to prevent, remove, compensate for or minimize, as appropriate, the adverse effects of activities or obstacles that seriously impede or prevent the migration of the species; and
> (c) to the extent feasible and appropriate, to prevent, reduce or control factors that are endangering or are likely to further endanger the species, including strictly controlling the introduction of, or controlling or eliminating, already introduced exotic species.

[9] Diario Oficial de la República de Chile, December 12, 1981.

[10] Sistema peruano de información jurídica https://spij.minjus.gob.pe/spij-ext-web/#/detallenorma/H773669.

[11] CMS Article I (1)(a).

[12] CMS Artícle II (1).

[13] CMS Artícle II (3)(b).

[14] CMS Artícle III (4).

CMS also establishes that it is prohibited to remove migratory species listed in Appendix I from their natural environment, with specific exceptions to this obligation.[15]

The marine otter was included in Appendix I during the Fifth Conference of the Parties (COP 5), held in Genova from April 10 to 16, 1997. The proponent for this inclusion was the Government of the Argentine Republic. It was argued that this species is migratory according to the Convention's definition since it would be "regularly crossing the national borders of the southernmost part of Chile and Argentina throughout its annual cycle, more specifically along the southeastern coast of the Fuegian archipelago, between Cape Horn and Isla de Los Estados, an extension of about 150 km. Additionally, there may be movements on the border between Chile and Peru."[16]

The most recent international call for attention on the importance and urgency of implementing conservation measures for otters worldwide occurred during the IUCN World Conservation Congress in Marseille in 2020.[17] The motion "Saving the world's otters" was presented and approved. The motion:

1. URGES Members, otter range states and other stakeholders to support the goals and objectives of the IUCN/SSC Global Otter Conservation Strategy and other efforts to address threats to otters by:
   (a) maintaining and enhancing otter habitats and eliminating illegal and unsustainable capture or killing of live otters;
   (b) developing and applying national wildlife protection legislation and international law to protect otters, including by monitoring, regulating or prohibiting their capture and sale for local and international trade;
   (c) eliminating the illegal trade in otters by increasing law-enforcement effectiveness in range and consumer countries, ensuring compliance with obligations and national regulatory frameworks, and reducing market demand for illegally collected otters or their parts and products;
   (d) engaging in scientific research and otter population surveys, as well as educational and awareness activities;
   (e) increasing societal support for otters and their environment and facilitating peaceful co-existence between otters and people;
   (f) ensuring that the management of all captive-bred otters is integrated with conservation interventions for wild populations; and
   (g) providing funding for activities outlined in the SSC Otter Specialist Group's Global Otter Conservation Strategy.

[15] CMS Article III (5).

[16] CMS COP5 Propuesta I/2, PROPUESTA DE INCLUSION DE ESPECIES EN LOS APENDICES DE LA CONVENCION SOBRE LA CONSERVACION DE LAS ESPECIES MIGRATORIAS DE ANIMALES SILVESTRES.

[17] https://www.iucncongress2020.org/es/motion/114.

7.2.2 *Legal Framework for the Conservation of the Marine Otter in Chile*

On June 18, 1929, the Hunting Law "Ley N° 4.601 Establece las disposiciones porque se regirá la caza en el Territorio de la República" was enacted[18] marking the first of its kind in Latin America (Servicio Agrícola y Ganadero, 2015). This general regulation outlines the procedures of capture, as well as penalties. Concerning marine otters, the following is specified:

> Article 6. – The export of hides or skins of the species indicated below shall be taxed as follows:
>
> ...
>
> Southern river otters or chungungos, 15 pesos per gross kilogram.
>
> Article 12. Those who hunt, buy, sell, or transport game, trade, or destroy eggs or offspring of animals during closed seasons or violate any other provision of this law shall be punished with imprisonment from one to sixty days, commutable to a fine of ten to five hundred pesos.
>
> The same penalty shall be imposed on those who, during closed seasons, sell, buy, or transport hides or skins of game animals captured within the closed season.[19]

In the same year, on November 15, 1929, Decree N° 4844 was enacted 'Approving the Regulation of the Hunting Law,' and it was published on December 21 of the same year.[20] This regulation addresses, among other matters, hunting methods, penalties, and prohibitions, including restrictions on the number of captured species and indefinite or temporary closed seasons. Regarding marine otters, the first closed season of 3 years is established:

> Article 2. The hunting of the following species is prohibited for the specified time:
>
> ...
>
> b) For three years: Different species of foxes, chingues, Southern river otters, and chungungo or sea cat.
>
> ...

[18] Diario Oficial de la República de Chile, July 1, 1929.

[19]

> Artículo 6°. – La exportación de cueros o pieles de las especies que se indican a continuación, estará gravada en la forma siguiente:
>
> ...
>
> Huillines o chungungos, 15 pesos kilo bruto.

Artículo 12. Los que cacen, compren, vendan o transporten piezas de caza, comercien o destruyan los huevos o crías de los animales en los períodos de veda o infrinjan cualquier otra disposición de la presente ley, serán penados con prisión de uno a sesenta días, conmutables en multa de diez a quinientos pesos.

Igual pena sufrirán los que, dentro de los períodos de veda, vendan, compren, o transporten pieles o cueros de animales de caza capturados dentro de la veda.

[20] Decreto N° 4844 que "Aprueba el Reglamento de la Ley sobre la Caza" Diario Oficial de la República de Chile, December 21, 1929.

The prohibitions established in subsection b) shall come into effect on December 1 of this year.[21]

Supreme Decree No. 531 of 1966 established an indefinite prohibition throughout Chile for the hunting of the two otter species, chungungo and huillín.[22] Subsequently, Supreme Decree No. 506 of Agriculture in 1967 repealed Supreme Decree No. 531 and indefinitely prohibits hunting, transportation, possession, and commercialization of the chungungo throughout the territory of the Republic.[23]

Decree No. 40 of 1972 amends the Regulation of Law No. 4601 and indefinitely prohibits the hunting, transportation, commercialization, possession, and industrialization of the Chungungo or Sea Cat or Marine Otter (*Lutra felina*).[24]

On November 22, 1989, Law No. 18892, the General Fisheries and Aquaculture Law, was enacted[25] incorporating under its jurisdiction hydrobiological species, defined as any species of hydrobiological organism in any phase of its development, whose normal or more frequent habitat is in the water.

In 1993, Decree No. 133 was published, a new regulation of the Hunting Law No. 4601,[26] in which a 20-year conservation closed season was established for all vertebrate animals, except those specified. The marine otter was not among them. However, Article No. 40 states:

Article 40 – The National Fisheries Service may authorize the capture of the following species of wild animals that are common in the marine environment, without prejudice to the

---

21

Art. 2. Prohíbese la caza de las siguientes especies por el tiempo que se indica:

...

b) Por tres años: Las distintas especies de zorros, chingues, huillín y chungungo o gato de mar.

...

Las prohibiciones establecidas en el inciso b) empezarán a regir desde el 1ro de diciembre del presente año.

22 Decreto N° 531 REEMPLAZA LOS ARTICULOS 1° Y 2° DEL DECRETO SECCION SEGUNDA N° 4844 DE 15 DE NOVIEMBRE DE 1929, EXPEDIDO POR EL EX MINISTERIO DE FOMENTO QUE REGLAMENTA LA LEY N°4.601 SOBRE CAZA Diario Oficial de la República de Chile, October 22, 1966.

23 Decreto N° 506 DEROGA EL DECRETO SUPREMO N° 531, DE 30 DE SEPTIEMBRE DE 1966, Y REEMPLAZA LOS ARTS. 1°, 2° Y 5° DEL DECRETO SECCION SEGUNDA N° 4.844, DE 15 DE NOVIEMBRE DE 1929, EXPEDIDO POR EL EX MINISTERIO DE FOMENTO, QUE REGLAMENTA LA LEY N° 4.601, SOBRE CAZA. Diario Oficial de la República de Chile, August 26, 1967.

24 Decreto N° 40 MINISTERIO DE AGRICULTURA, DEROGA LOS DECRETOS QUE INDICA Y MODIFICA EN LA FORMA QUE SEÑALA EL REGLAMENTO DE LA LEY 4.601, SOBRE CAZA, Diario Oficial de la República de Chile, march 9, 1972.

25 Ley N° 18,892 Ley General de Pesca y Acuicultura, Ministerio de Economía FOMENTO Y RECONSTRUCCIÓN, Diario Oficial de la República de Chile, December 23, 1989.

26 Decreto N° 133 Ministerio de Agricultura, APRUEBA REFLAMENTO DE LA LEY DE CAZA, Diario Oficial de la República de Chile, March 9, 1993.

provisions of Law No. 18,892 on Fisheries and Aquaculture and the International Conventions signed by the country for the protection of these species.

...

CARNIVORA ORDER

Mustelidae Family, only the genus Lutra Otters[27]

The Law No. 19473, which replaces Law No. 4601 on hunting, exempts its application to hydrobiological species and resources, stating that their preservation is governed by Law No. 18892, the General Fisheries and Aquaculture Law.[28] It also establishes that the species or groups of mammal species constituting hydrobiological species or resources will be determined by the regulations of this law. This is done through Decree No. 5 of the Ministry of Agriculture in 1998, which approves the regulation of Law No. 19473[29] considering the Mustelidae family, including otters and huillines, only of the genus Lontra Marine otters then fall under the category of "Hydrobiological Resources" and are under the jurisdiction of the Ministry of Economy, Development, and Reconstruction, subject to absolute protection along the entire coastline of the country.

Supreme Decree No. 225, published on November 11, 1995,[30] with modifications in 2005 and 2007,[31] must also be mentioned. This decree establishes a 30-year extraction ban for some marine vertebrate species, including the marine otter or chungungo.

The Red Book of Terrestrial Vertebrates of Chile (Glade, 1988) categorized the marine otter as a Vulnerable species. On March 24, 2007, Decree No. 151 of the Ministry of the General Secretariat of the Presidency was published, classifying the Chungungo *Lontra felina* as "Insufficiently Known".[32]

27

Artículo 40 – El Servicio Nacional de Pesca podrá autorizar la captura de la captura de las siguientes especies de animales silvestres que son habituales del medio marino, sin perjuicio de lo establecido en la ley N° 18.892 sobre Pesca y Acuicultura y las Convenciones Internacionales suscritas por el país para la protección de estas especies.

...

ORDEN CARNIVORA

Familia Mustelidade, solo el género Lutra Nutrias.

28 Ley N°19,473 SUSTITUYE TEXTO DE LA LEY N° 4.601, SOBRE CAZA, Y ARTICULO 609 DEL CODIGO CIVIL, Diario Oficial de la República de Chile, September 27, 1996.

29 Decreto N° 5 MINISTERIO DE AGRICULTURA, APRUEBA REGLAMENTO DE LA LEY DE CAZA, Diario Oficial de la República de Chile, December 7, 1998.

30 Decreto N° 225 ESTABLECE VEDA PARA LOS RECURSOS HIDROBIOLOGICOS QUE INDICA, EXPEDIDO POR Ministerio de Economía, Fomento y Reconstrucción, Diario Oficial de la República de Chile, November 11, 1995.

31 Decreto N°135 Ministerio de Economía, Fomento y Reconstrucción, Diario Oficial de la República de Chile, 26 de enero de 2005 y Decreto N° 434 Ministerio de Economía, Fomento y Reconstrucción, Diario Oficial de la República de Chile, February 16, 2007.

32 Decreto N° 151 Ministerio Secretaría General de la Presidencia, OFICIALIZA PRIMERA CLASIFICACION DE ESPECIES SILVESTRES SEGÚN SU ESTADO DE CONSERVACION, Diario Oficial de la República de Chile, March 24, 2007.

Until the year 2021, the marine otter was classified in Chile as **ENDANGERED (EN) A3ce**, according to the provisions of Decree No. 44 of the Ministry of the Environment.[33] Finally, in the current Fisheries and Aquaculture Law, there are regulations regarding distancing for the observation of marine mammals and other provisions related to the species (Subsecretaría de Pesca y Acuicultura, 2018).

Especially noteworthy is what is established in the Political Constitution of the Republic, particularly in Article 19 No. 8: "The right to live in an environment free of pollution. It is the duty of the State to ensure that this right is not affected and to protect the preservation of nature. The law may establish specific restrictions on the exercise of certain rights or freedoms to protect the environment."[34]

7.2.3 Legal Framework for the Conservation of the Marine Otter in Peru

Until the 1940s in Peru, there were purportedly no conservation policies for renewable natural resources, and the state did not coordinate any concerted efforts in this regard, except for isolated actions (Ponce, 1973). Concerning the marine otter, it wasn't until 1951 that this species was mentioned in Supreme Resolution No. 236. This resolution declared a closed season and prohibited hunting in the Coastal Region from December 15 to March 31 of each year for the marine otter (*Lutra felina*), among other species.[35] The regulation also stipulated that offenders would face fines ranging from 50 to 500 soles oro, imprisonment, and the confiscation of hunted animals, their skins or hides, and their weapons.

In 1967, the Agricultural Promotion and Development Law No. 16726 was enacted.[36] Within its Chapter VI, titled "Protection of Wildlife," it mandated that the Forest and Game Service would define rational prohibitions and closed seasons for hunting, fishing, and the export of live animals and skins to prevent their complete extinction.

[33] Decreto Número 44 Ministerio del Medio Ambiente, APRUEBA Y OFICIALIZA CLASIFICACIÓN DE ESPECIES SEGÚN ESTADO DE CONSERVACIÓN, DECIMOSÉPTIMO PROCESO, Diario Oficial de la República de Chile, December 30, 2021.

[34] "El derecho a vivir en un medio ambiente libre de contaminación. Es deber del Estado velar para que este derecho no sea afectado y tutelar la preservación de la naturaleza. La ley podrá establecer restricciones específicas al ejercicio de determinados derechos o libertades para proteger el medio ambiente" Decreto 100. Texto Refundido, Coordinado y Sistematizado de la Constitución Política de la República de Chile. https://www.bcn.cl/leychile/navegar?idNorma=242302.

[35] Resolución Suprema No. 236 June 4, 1951, Declara épocas de veda y prohíbe caza.

[36] Ley N° 16,726 Declarando de Interés Social y de Necesidad Nacional la Promoción y Desarrollo del Sector Agropecuario del País, November 16, 1967.

In 1975, Peru ratified the Convention on International Trade in Endangered Species of Wild Fauna and Flora (CITES), thereby committing to the obligations outlined in the Convention.

Decree Law No. 21147, the Forest and Wildlife Law of 1975, regulated the conservation of wildlife. It mandated that the Executive Branch could indefinitely prohibit the exploitation of species facing extinction. The Regulation for this Law, Supreme Decree No. 158-77-AG[37] outlined the various categories that wildlife species could be classified into based on their conservation needs. It also stipulated that a Ministerial Resolution would establish the list of protected species.

Ministerial Resolution No. 01710-77-AG-DGFF marked the initial official classification of endangered and threatened species. Within this resolution, the marine otter (*Lutra felina*) was categorized as a "Species in Danger of Extinction".[38] Later, in the second classification in 1990, through Ministerial Resolution No. 01082-90-AG/DGFF, the marine otter (*Lutra felina*) is once again placed in this category (Pulido, 1991).

A transformative shift occurred with the enactment of the new Constitution in 1993. Elevating the commitment to biodiversity conservation to a constitutional mandate, Peru incorporated the obligation of the state to promote the conservation of biological diversity and protected natural areas in Article 68 of the Constitution of Peru (1993).

Within this constitutional framework, Peru became a party to the Convention on Biological Diversity through ratification via Legislative Resolution No. 26181 on April 30, 1993. Moreover, Article 7 of Law No. 26839, the law on the conservation and sustainable use of biological diversity, stipulates that the National Biodiversity Strategy serves as the primary planning instrument for fulfilling the objectives of the law and the Convention.

In 1997, Peru acceded to the Convention on the Conservation of Migratory Species of Wild Animals (CMS). In 2004, through Supreme Decree No. 034-2004-AG, the categorization of threatened species of wildlife was approved, imposing a ban on their hunting, capture, possession, transportation, or commercial export. This decree, in its annex, classified the marine otter (*Lontra felina*) under the "Endangered" category. Article 4 of this decree, titled "Scientific Collections of Critically Endangered and Endangered Species," authorized the hunting or scientific collection of specimens threatened with extinction (categorized as Critically Endangered or Endangered). However, such activities were permitted only when research contributed to the conservation of these species and served the national interest and benefit. Approval from the National Institute of Natural Resources (INRENA), the competent environmental authority at that time, and recognized

[37] Decreto Supremo N°158-77-AG REGLAMENTO DE CONSERVACION DE FLORA Y FAUNA SILVESTRE, March 31, 1977.

[38] Resolución Ministerial N° 01710-77-AG-DGFF Aprueban clasificación de Flora y Fauna Silvestre y fijan precios por ejemplar para la extracción comercial de productos de la Fauna, September 30, 1977.

Lutra felina **Especie en vía de Extinción**.

national or international scientific institutions was mandatory, as deemed appropriate by the case.

Subsequently, through Supreme Decree No. 004-2014-MINAGRI, the update of the aforementioned list was approved. The marine otter continues to be categorized as "Endangered" (EN). Article 7 maintains the authorization for scientific collection for research purposes, specifying that for species categorized as "Critically Endangered" (CR) and "Endangered" (EN), "the collection is authorized when specimens are not available in the collections of natural history museums or other scientific institutions accredited by the Ministry of Agriculture and Irrigation. Exceptionally, this provision does not apply to the collection of biological samples, which must follow protocols established by the National Forestry and Wildlife Authority. These protocols should consider current national and international animal welfare criteria." Additionally, Article 10 outlines the final disposition of found or confiscated specimens, stating that "the final disposition of dead or alive specimens, found or confiscated, their parts, and derivatives belonging to threatened species, is the responsibility of the National Forestry and Wildlife Authority".[39]

Additionally, according to the current Forestry Law, Law No. 29763, Forestry and Wildlife Law, Article 6 states that "wildlife resources are non-domesticated animal species, native or exotic, including their genetic diversity, that live freely in the national territory, as well as individuals of domesticated species that, due to abandonment or other causes, assimilate to wildlife habits, except species different from amphibians born in marine and continental waters, which are governed by their own laws. The scope of this law includes wildlife specimens (living or dead specimens, eggs, and any part or derivative), individuals kept in captivity, as well as their products and services."

In addition to the mentioned regulations, there is the General Fisheries Law, approved by Legislative Decree No. 25977, which, in Article 151 of its Regulation, approved by Supreme Decree No. 012-2012-PE, states that "hydrobiological resources are animal and plant species that develop all or part of their life cycle in the aquatic environment and are susceptible to exploitation by humans." Under this definition, the marine otter would be considered as it belongs to the class of marine mammals. However, as it is not susceptible to exploitation by humans, it would no longer fall within the scope of the defined term. This formal situation is not a minor issue since it implies that the fishing sector cannot deploy specific actions for its protection.

The legal framework for the conservation of the marine otter in Peru has evolved over the years, with various regulations, laws, and international agreements coming into play. The country has committed to protecting endangered species through its participation in conventions such as CITES and CMS. The constitutional and legislative provisions emphasize the need for biodiversity conservation and the protection of natural habitats. However, challenges persist in ensuring effective

[39] Supreme Decree 004-2014-MINAGRI includes a classification and categorization list of threatened species. The categories are: Critically Endangered (CR), Endangered (EN), Vulnerable (VU), Near Threatened (NT), and Data Deficient (DD).

enforcement and coordination among various agencies to safeguard the marine otter's population and habitat.

7.3 Judicial and Administrative Actions in Defense and Conservation of the Marine Otter

For the purpose of the relevant legal and administrative actions, examples from Chile will be presented, ranging from targeted captures to processes related to infrastructure construction.

In the Peruvian case, we will primarily highlight administrative and legal actions in response to the 2022 oil spill along the Ventanilla coast (near Lima), where the marine otter has been one of the affected species.

7.3.1 *Cases in Chile*

Despite existing legal prohibitions, marine otters have been hunted by artisanal fishermen, and accidental captures have occurred in nets and traps in some coastal areas of Chile (Olavaria et al., 2007). As a result, a complaint was filed that led to the first nationwide trial for the death of marine otters in 2003. The process concluded with a guilty verdict for the death of marine otters in shrimp traps in the commune of Puchuncaví.[40]

In 2008, a Protective Action was filed against the Environmental Qualification Resolution approved by the State Services for the construction and operation of the Campiche Thermoelectric Plant, due to the severe environmental consequences in the Quintero Bay, which is declared a Metal and Sulfur Dioxide (SO_2) Saturated Zone. A process was initiated and ruled upon by the Court of Appeals of Valparaíso. In its verdict, Cause Rol 317-2008, it states: "a protective action is filed against the Regional Commission of the Environment of Valparaíso, represented by its President … and against the public services that make it up, domiciled at Melgarejo 669 floor 19 and Avda. Pedro Montt 1992, Valparaíso, due to the issuance of Exempt Resolution No. 449, of May 9, 2008, notified to that party on the 22nd of the same month and year, which favorably qualified the Environmental Impact Study of the Campiche Thermoelectric Project, whose owner is the company Eléctrica Campiche S.A., without complying with the legal and regulatory provisions in effect, as well as due to lack of oversight, acts, and omissions that violate the guarantee enshrined in Article 19 No. 1, 8, 9, and 21, that is the right to life, the right to health protection, the right to live in an environment free of pollution since it has not considered the involved environmental conditions, and this, considering that it also lacks the

[40] Appendix 1.

technical qualification of the case. Consequently, it constitutes a threat or attack against the balance of the environmental system, caution, or reservation that is part of the preventive work that the State must safeguard in this matter to avoid or prevent the inconveniences or damages that may be feared. This is a fundamental environmental principle emanating from international sources called the precautionary principle, which states that when there is a threat of serious and irreversible harm to nature, the lack of absolute scientific certainty should not be used as an excuse to postpone the adoption of effective measures to prevent environmental degradation. This principle is also recognized in the Message of the Environmental Framework Law when it states that the preventive principle is embodied in the law in four actions, indicating among them the Environmental Impact System. Therefore, this protective action must be upheld, since the appealed resolution is illegal and threatens a fundamental guarantee, which is the right to live in an environment free of pollution."

The entry of marine otters into the cooling system of these thermoelectric plants, in addition to the destruction of the habitat due to the intensive use of the coastal edge, are the main threats to the species in Quintero Bay (IFOP, 2016).

Another relevant legal action was the filing of a criminal complaint before the Court of Letters and Guarantee of Quintero, for animal abuse and the death of an adult male marine otter, which was suctioned into the seawater cooling duct of the AES Corp. Thermoelectric Complex, which occurred in the intake wells on September 5, 2018 (Figs. 7.1, 7.2 and 7.3).[41]

As is commonly done, a multisectoral legal and administrative strategy was developed to address the events that occurred within the facilities of the Thermoelectric Plant. The information was obtained from the company itself, and a series of administrative actions, complaints, and information were activated for the following State services: (a) Superintendence of the Environment, (b) Undersecretariat of Fisheries and Aquaculture, (c) National Fisheries and Aquaculture Service, (d) Chilean Navy, as well as criminal actions through a Complaint to the Prosecutor's Office of Quintero and through a Criminal Complaint to the Criminal Guarantee Court of Quintero. The Court declared the definitive dismissal of the case because the facts were not constitutive of a crime.

Another case involved an oil spill in Quintero Bay caused by the B/T Mimosa, during towing by an offshore tug on September 24, 2014, at the ENAP Monobuoy Maritime Terminal (ENAP Refinerías S.A., 2019). An Action for declaration and reparation of environmental damage was filed. Among the reported environmental damages was the death of a marine otter (chungungo). On March 13, 2018, the Second Environmental Court of Santiago rejected the claim for environmental damage reparation, stating that there was no evidence in the file to prove the existence of significant damage to the environment. The Municipality of Quintero appealed

[41] Rol: N° Ordinaria-5-2019. Juzgado de Letras y Garantía de Quintero, January 3, 2019 https://oficinajudicialvirtual.pjud.cl/indexN.php#.

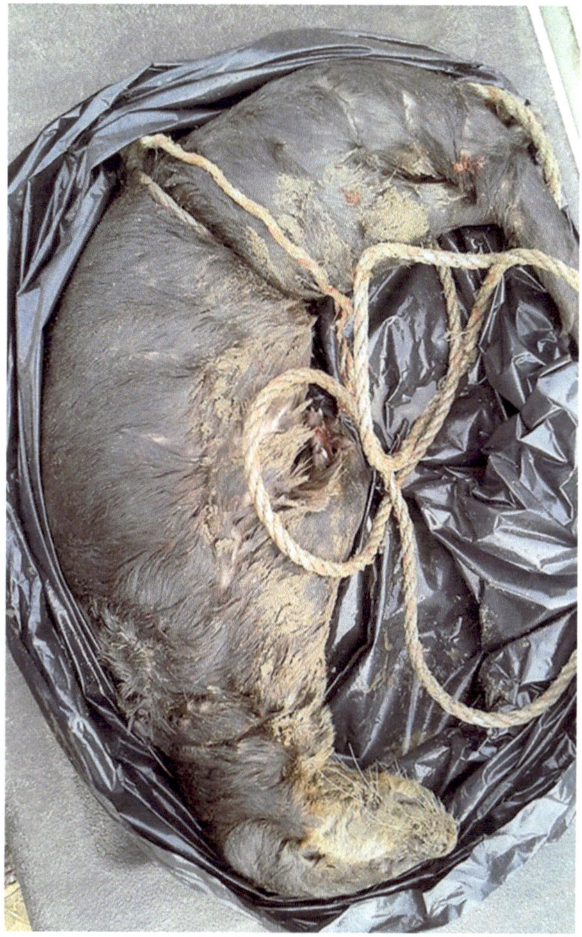

Fig. 7.1 Marine otter trapped and dead in suction ducts of the AES-GENER company, September 2018 in Bahía de Quintero, Chile. (Source: Servicio de Pesca y Acuicultura Región de Valparaíso, Chile)

the previous judgment on April 3, 2018, before the Supreme Court, through a cassation appeal (ENAP Refinerías S.A., 2019).

On September 25, 2019, the Supreme Court issued a ruling accepting the cassation appeal and, consequently, a replacement ruling declaring the existence of environmental damage[42] (ENAP Refinerías S.A., 2019). The judgment orders ENAP Refinerías S.A. to implement precautionary measures, including reporting to the Ministry of the Environment. These measures, coordinated with CONAF, included the implementation of a "monitoring plan to identify, quantify, and preserve the development conditions of the Humboldt penguin, common tern, chungungos, and other flora, fauna, and biota affected by the oil spill." This judgment has been final and enforceable since October 9, 2019 (ENAP Refinerías S.A., 2019) (Fig. 7.4).

[42] Rol: N° 13.177-2018 https://tribunalambiental.cl/wp-content/uploads/2020/03/CS_13177-2018_2TA_D-13-2014_Sentencia.pdf.

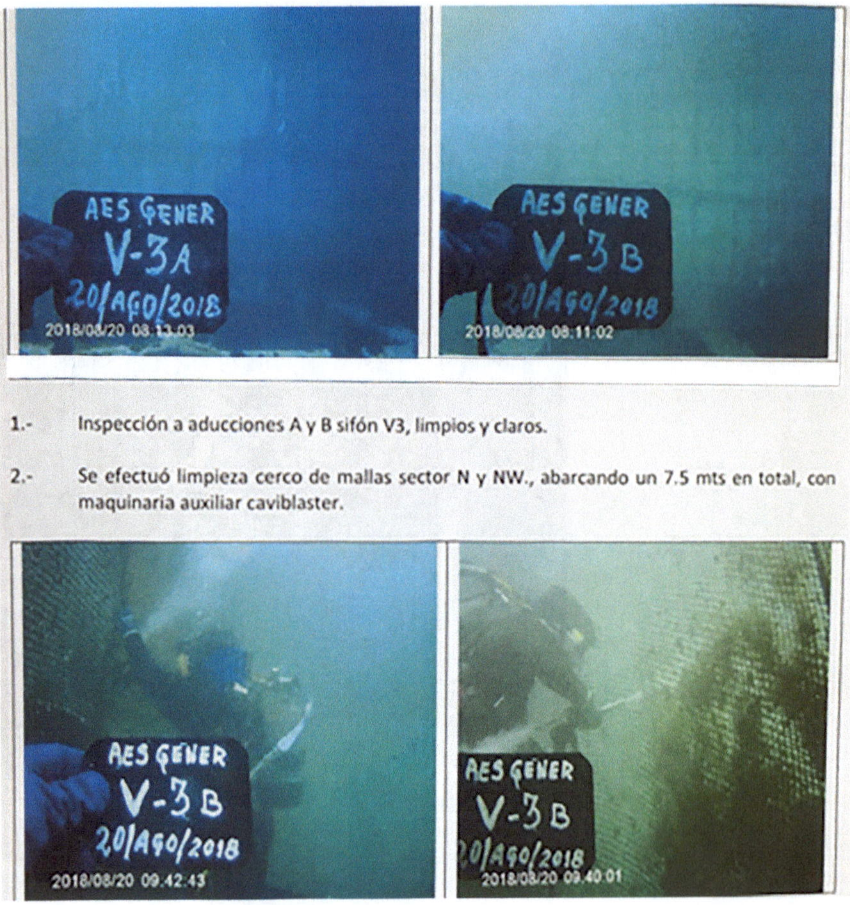

Fig. 7.2 Underwater images depicting the grates in the seawater intake siphons of the company AES-GENER. (Source: Informe Fiscalización Superintendencia de Medio Ambiente Diciembre 2019. Informe Fiscalización Muerte Nutria RCD.docx)

7.3.2 *Cases in Peru*

In 2022, the largest oil spill occurred on the sea coast of Lima (in the town of Ventanilla), during the unloading operations of the Mare Doricum Tanker, “on Saturday, January 15, 2022, at 10:26 p.m., Refinería La Pampilla S.A.A., (…), reported an environmental emergency to the Environmental Evaluation and Enforcement Agency (OEFA) due to the presence of an oily substance in a 2.5 m2 area by the sea, indicating a possible spill of 0.16 barrels of hydrocarbon. OEFA verified the presence of hydrocarbon in the sea and on the beach strip (sand). The National Service of Natural Areas Protected by the State (SERNANP) confirmed that the spilled oil was moving north, affecting the areas of the Fishing Group Islets

Fig. 7.3 Marine otter trapped in the intake well of the cooling process at AES GENER company. (Source: Servicio de Pesca y Acuicultura Región de Valparaíso, Chile)

of the National Reserve System of Islands, Islets, and Guano Points in 512 Ha, and the Reserved Zone Ancón in 1758.1 Ha, impacting the biodiversity of the area. (sic)".[43] Later, the Ministry of the Environment specified the spill volume at 11,900 barrels, and on January 25, an additional spill of 8 barrels was reported at the same station.[44] It is worth noting that competencies related to wildlife in this case fall under SERNANP inside protected natural areas according to Legislative Decree No.

[43] Informe N° 00009-2022-MINAM/VMGA/DGCA de la Dirección General de Calidad Ambiental del MINAM.

[44] https://rpp.pe/politica/gobierno/minam-estima-en-11-900-barriles-el-derrame-de-petroleo-casi-el-doble-que-el-reportado-por-repsol-noticia-1383456.

Fig. 7.4 Authorities and cleaning staff of the ENAP spill on Loncura beach, Bahía de Quintero, 2014, Valparaíso, Chile, 2014. (Source: Ricardo Gonzalo Correa)

1079,[45] while outside these areas, they belong to the National Forest and Wildlife Service (SERFOR).

In this regard, the Office of the Ombudsman issued a report stating, "By contaminating these ecosystems, the food sources of wildlife are also contaminated, including vulnerable or endangered species such as the South American sea lion and the fur seal, or the Humboldt penguin and the **marine otter**,[46] respectively. As a result, these specimens will become ill and possibly die, even without showing traces of oil on their bodies. While the OEFA issued a set of administrative measures[47] aimed at protecting protected natural areas and wildlife, these measures are not proving sufficient to achieve this goal. In this regard, the death of at least 5 of the 15 marine otters inhabiting these ecosystems has already been reported.[48] Given the circumstances, it is important to note that both SERFOR and SERNANP, as the governing

[45] Decreto Legislativo N° 1079 DECRETO LEGISLATIVO QUE ESTABLECE MEDIDAS QUE GARANTICEN EL PATRIMONIO DE LAS ÁREAS NATURALES PROTEGIDAS.

[46] Authors' emphasis.

[47] Resolución N° 007-2022-OEFA/DSEM.

[48] Sociedad Peruana de Derecho Ambiental. Estudio confirma disminución de población de nutrias tras derrame de petróleo. En: https://www.actualidadambiental.pe/estudio-confirma-disminucion-nutrias-derrame-repsol/.

bodies of the functional systems that integrate the competent authorities in these matters—SINAFOR[49] and SINANPE,[50] respectively—have the valuable task of taking necessary actions to protect the impacted biodiversity and ecosystems" (Defensoría del Pueblo, 2022).

In this intricate scenario, the Office of the Ombudsman underscored the absence of protocols to address environmental emergencies, such as the one described, and its repercussions on wildlife: "Indeed, from the review of the legal framework that regulates the wildlife sector, it is evident that there are no protocols for the development of activities such as rescue, rehabilitation, release, restocking, monitoring, and population assessment of wildlife." It is crucial to emphasize the necessity for protocols governing the release, and, if applicable, restocking, followed by thorough monitoring and evaluation in ecosystems to ensure the protection and conservation of the marine otter. This underscores the immediate need for comprehensive protocols designed to safeguard the marine otter in situations involving hazardous activities.

A lawsuit related to the impact on fishermen's collectives[51] has been initiated by the National Institute for the Defense of Competition and the Protection of Intellectual Property (INDECOPI), presented before the courts in Lima: "The 27th Specialized Civil Court of Lima, through resolution number 4, has accepted for processing the claim for compensation for damages filed by INDECOPI, seeking an amount of USD 4.5 billion (four billion five hundred million US dollars)". Multiple companies are named as defendants.

The Provincial Prosecutor's Office Specialized in Environmental Crimes of the Lima Northwest Judicial District (FEMA Lima Northwest) has initiated an investigation into the alleged crime of environmental pollution[52] against the legal representatives and officials of Refinería La Pampilla S.A.A., Grupo Repsol del Perú S.A.C., and those deemed responsible. The public attorney's office Specialized in Environmental Crimes has appeared in this case within the framework of the violated environmental interests that it represents.

Regarding the criminal complaint[53] for environmental damage due to the spill,[54] for its processing and success, two concurrent elements are required. First, incorporating the definition of environmental damage linked to liability for the exercise of

[49] Artículo 13 de la Ley Forestal y de Fauna Silvestre, Ley N° 29763.

[50] Artículo 8 de la Ley de Áreas Naturales Protegidas, Ley N° 26834.

[51] The legal action is focused on the personal sphere: "(…) the impact on diffuse interests with collective moral damage; we seek compensation for the affected population, third-party beneficiaries of the civil liability coverage residing within 150 km of the contaminated coastline." "(…) la afectación de los intereses difusos con daño moral colectivo; buscamos el resarcimiento de la población afectada, terceros beneficiaros (sic) de la cobertura de responsabilidad civil que domicilian en 150 km del litoral contaminado". https://www.gob.pe/institucion/indecopi/noticias/.

[52] Under the crime of pollution as stipulated in Article 304 of the Peruvian Penal Code (Código Penal Peruano).

[53] The complaint, as of the cutoff date of October 6, 2022, has not been filed yet.

[54] Interview with the Attorney General's Office specializing in Environmental Matters.

a risky activity (strict liability),[55] where elements of analysis such as intention, knowledge, and causal link are not applicable, based on the premise that environmental damages are characterized by uncertainty. Second, including in the evidence a single damages report from competent administrative entities, incorporating what has been done based on the reports issued by these entities regarding the impacts on abiotic and biotic factors from the spill, and including the economic assessment and quantification of the damage, carried out by the Directorate of Valuation and Expertise of the Attorney General's Office, to demand compensation for pure ecological damage to the sea, wildlife, including the marine otter,[56] and the payment of compensation for the violation and impact on ecosystem services associated with natural resources.

7.4 Legal Standard and Regulations Needed to Enhance Species Conservation Management

The ongoing mortality of individuals in Quintero Bay (Chile), particularly in the pipelines, siphons, and seawater intake wells for the industrial processes of multiple port companies, has prompted questioning, scrutiny, intervention, and the need to update the following issues:

---

[55] In this regard, Peruvian regulations limit "environmental damage," understanding the latter as both biotic and abiotic components, but also with some utility, benefit, or use for humans. This restriction poses a limitation on the protection scope for species such as the marine otter. Article 142 Ley 28611, Ley General del Ambiente: "142.2. Responsabilidad por Daños Ambientales. Se denomina daño ambiental a todo menoscabo material que sufre el ambiente y/o alguno de sus componentes, que puede ser causado contraviniendo o no disposición jurídica, y que genera efectos negativos actuales o potenciales."("142.2. Environmental Damage Liability. Environmental damage is defined as any material harm suffered by the environment and/or any of its components, whether caused in violation of legal provisions or not, and that produces current or potential negative effects."); also, see the Constitutional Court Judgment 048-2004-TC: "27. (…) The environment is defined as "(…) the set of social, cultural, biotic, and abiotic elements that interact in a specific space and time; which could be depicted as the sum of nature and human manifestations in a specific place and time. The term 'biotic' refers to all living beings in the same region that coexist and influence each other; whereas 'abiotic' refers to non-living things, such as water, air, subsurface, etc. The environment consists of the so-called natural elements, which, depending on the case, can generate some type of utility, benefit, or use for human existence or coexistence." (Sentencia del Tribunal Constitucional 048-2004-TC "27.(…) El medio ambiente se define como "(…) el conjunto de elementos sociales, culturales, bióticos y abióticos que interactúa en un espacio y tiempo determinado; lo cual podría graficarse como la sumatoria de la naturaleza y las manifestaciones humanas en un lugar y tiempo concretos… El término biótico se refiere a todos los seres vivos de una misma región, que coexisten y se influyen entre sí; en cambio lo abiótico alude a lo no viviente, como el agua, el aire, el subsuelo, etc. El medio ambiente se compone de los denominados elementos naturales, los cuales pueden generar, según sea el caso, algún tipo de utilidad, beneficio o aprovechamiento para la existencia o coexistencia humana.").

[56] The marine otter is an indicator of the good health of the oceans, https://sinia.minam.gob.pe/documentos/conociendo-indice-salud-oceano.

(a) Regulations for the construction of seawater intake structures, with the aim of mitigating biodiversity loss, especially of marine otters.
(b) Protocols for the rescue of marine otters, to adapt them to new technologies, methods, and tools for better performance in this complex task.
(c) Induction processes for coastal project workers, aimed at raising awareness of the presence of marine otters.
(d) Contributions to new remote observation methodologies, to identify active burrows and determine the presence of the species in intervened or to-be-intervened locations.
(e) Standards, ordinances, and guidelines for good practices in coastal areas with the presence of marine otters.

In the Peruvian case, in response to the oil spill in the Ventanilla Sea and the subsequent death of marine otters, attention is focused on addressing two legal aspects. Firstly, reformulating the definition of environmental damage to access effective jurisdictional protection; and secondly, the need for coordination among entities with competencies in the matter, as administrative sanctioning procedures have been initiated depending on the exercise of their functions.[57] Authorities operate within a dense legal framework, facing the challenge of formal and effective coordination to achieve solutions for the environment and affected species. In this regard, "the management of coastal areas, specifically the littoral, requires coordinated authority," and "it should not be overlooked that each sector is the environmental authority in matters within its competence. This has led, especially in terms of environmental protection and sanction, to delays in the respective procedures and formalities, to the detriment of effective protection"[58] (Garay, 2003).

Therefore, achieving the necessary protection against the problems and threats affecting the marine otter, such as illegal fishing with explosives, marine pollution, the presence of microplastics in bays, as well as illegal occupations of marine areas and human activities, requires not only the application of existing legal frameworks for precautionary and preventive protection but also substantive modifications to the conceptualization of environmental damage that incorporate strict ecological damage (including wildlife, the sea, and flora). Additionally, it is essential to establish a legal precedent through a judicial ruling that sets a precedent and includes compensatory payments for the damage to associated ecosystem services, aiming to achieve administrative and criminal penalties.

Another element limiting effective jurisdictional protection is that, according to Peruvian regulations, obligations arising from the environmental management

[57] There would be administrative sanctioning procedures initiated by Servicio Nacional de Áreas Naturales Protegidas (SERNANP), el Organismo Supervisor de la Inversión en Energía y Minería (OSINERGMIN), Organismo de Evaluación y Fiscalización Ambiental (OEFA), Dirección de Capitanías de Puerto (DICAPI) and Servicio Forestal y de Fauna Silvestre (SERFOR).

[58] Garay (2003, pp. 3, 181).

instrument are subject to inspection. Therefore, if specific actions regarding wildlife, such as the marine otter, are not expressly stated within the environmental impact assessment and its Contingency Plan, the requirement for supervision and penalties is reduced, absent, or null. This makes it imperative to have protocols for the species included in these environmental certification procedures.

7.5 Necessary and Urgent Measures for the Conservation of the Marine Otter

In the context of dwindling populations of the marine otter, the following technical measures are deemed necessary and urgent for its conservation:

1. Strengthening programs on understanding the behavior and characteristics of the species, as well as the ecosystem resources that sustain it, by inspectors and competent authorities.
2. Creating legal capacities to intervene in coastal projects subject to environmental assessment at the Chilean Environmental Evaluation Service. Review of current legislation, updating it to the requirements of the current conservation status with an emphasis on the precautionary principle enshrined in the law.
3. Promoting social and cultural campaigns in coastal communities to value the presence of the marine otter, as the "guardian" species in the defense of coastal areas in general, highlighting the importance of these actions for its conservation.
4. Development of a computational registry at the national and regional levels regarding places with the presence of the species, allowing digital updates with the support of competent authorities, organized coastal communities, research centers, and/or universities annually.
5. Generating national-level monitoring plans and programs for the species to establish the state of populations, such as relative density, through new observation technologies and reliable, innovative, and standardized methods.
6. Achieving, as soon as possible, the declaration of Multiple-use Marine Protected Areas, with the main focus on the conservation of the *Lontra felina* species and the ecosystems that support it.
7. Determining and prioritizing the main current threats to the conservation and preservation of the species in the intertidal coastal zones it inhabits.
8. Defining scientific studies, prioritizing those "in situ" that allow better knowledge of the species, its direct and indirect threats, and the vulnerabilities of its territories, for decision-making.
9. Development of Recovery, Conservation, and Management Plans for marine otters, addressing their current conservation status in Chile.

10. Having protocols from the State for the development of activities such as rescue, rehabilitation, release, repopulation, monitoring, and population assessment of the marine otter.
11. Incorporating protocols for environmental emergencies by companies in their environmental management instruments regarding the treatment of marine otters.
12. Including the main lines of action for species protection from the IUCN[59]:
    (a) Photos or records of live or dead animals and their causes, signs, visually documented records, and other credible evidence.
    (b) Development and promotion of a real-time global citizen science report system for otter records and information on trade and other threats.
    (c) Identifying the types and intensities of threats at the level of their populations (including, but not limited to, overexploitation, conflict between otters and humans, pollution, decrease in their prey, habitat degradation and fragmentation, climate change, invasive species, feral dogs, incidental death, and diseases) of globally or regionally important otters, recognizing that these will vary between coastal territories as part of community-based conservation.
    (d) Developing a precise understanding and sensitivity at the local level so that community-based interventions can work to ensure their development.

[59] Duplaix and Savage (2018). OSG, Action lines.

Appendix: First Judicial Ruling for the Death of a Marine Otter in Chile

Puchuncaví, a diecinueve de junio de dos mil tres.

Vistos:

1.- Que a fs. 1 consta la denuncia interpuesta por doña Nelly Figueroa Jeldres, bióloga marino, con domicilio en Normandie 1893 Of. 3 y 4 Quintero; en contra de cuidador domiciliado en Hotel Las Rocas s/n Maitencillo, por haber dado maltrato y muerte de un animal denominado Chungungo o Gato de Mar (Lutra felina) especie hidrobiológica protegida. Carabineros al constituirse en el domicilio del denunciado se percató que la especie marina se encontraba secando clavado en una tabla con su boca rellena con papel. Al ser interrogado manifestó libre y espontáneamente su autoría en los hechos, agregando que este animal había sido atrapado en una jaula para pescar jaibas y lo había encontrado muerto. Esta especie marina se encuentra protegida en el DS. 225-95 del 9 de Noviembre del 1995 (MINECOM)

2.- Que a fs. 2 se agrega copia del oficio de Carabineros dirigido al Servicio Nacional de Pesca donde consta haberse remitido la piel del Chungungo o Gato de Mar clavado en una tabla.

3.- Que a fs. 2 vta. consta de la Resolución del Juzgado de Letras de Quintero en virtud del cual se decreto la incompetencia del referido Tribunal.

4.- Que a fs. 4 se agrega la declaración indagatoria del denunciado Armando Vicencio Caicedo, ya individualizado quien reconoce su participación en los hechos, agrega que el mantenía una jaula en el mar para sacar jaibas, que al día siguiente al retirar las mismas se encontró en el interior que había un animal muerto que no conocía. Consultado con un pescador le manifestó que era un Chungungo, el cual le dijo que le convenía descuerarlo para poderlo secar.

5.- Que a fs. 9 y 1 se agrega la orden de investigación cumplida por Carabineros de Maitencillo, en la cual se estableció la veracidad del denuncio y la participación del denunciado. Además se deja constancia que el cuero del mamífero fue remitido al Servicio Nacional de Pesca de Quintero mediante el Oficio 54 de ese destacamento. Finalmente destaca que la especie en cuestión se encuentra protegida por D.S. 225 del 9 de Noviembre del año 1995, por un plazo de 30 años a contar de la fecha de su publicación.

6.- Que a fs. 15 de autos consta de la Resolución de este Tribunal que ordeno a las partes a comparecer con sus testigos y medios probatorios a la audiencia del día 20 de Febrero del 2003.

7.- Que a fs. 14 se hace parte el abogado don Ricardo Gonzalo Correa D. en representación del Grupo Ecológico "Chinchimen" de la comuna de Zapallar.

8.- Que a fs. 18 se deja constancia que a la audiencia antes indicada solamente compareció la parte denunciante, y que fue llevada a efecto en rebeldía del denunciado. Manifiesta la denunciante aparte de la multa que corresponde señala que el denunciado no tenia permiso alguno para extraer recursos marinos, según Ley que se adjunta.

9.- Que el Art. 14 de la Ley faculta para apreciar al Tribunal para apreciar los antecedentes de conformidad a las normas de sana critica, ello unido a la denuncia de Fs 1, a la declaración indagatoria de fs. 4 en la cual el denunciado reconoció su participación en los hechos, agregando que el animal lo descuero porque se encontraba muerto, la investigación de Carabineros de Maitencillo agregada a fs. 9 y 10 donde se constato la veracidad del denuncio, la presentación efectuada por el abogado Ricardo Gonzalo Correa D. de fs. 14, la audiencia de prueba de fs. 18 la que se llevo a efecto con la sola presencia de la parte denunciante de doña Nelly Figueroa Jeldres, y en rebeldía el denunciado; se encuentra a juicio de este Tribunal suficientemente acreditado el hecho denunciado y la participación del hechor al haber cometido actos de maltrato y la muerte del mamífero Chungungo o Gato de Mar, el cual no presento prueba de descargo alguno, habiendo infringido de esta forma lo dispuesto en el Art. Nº 295-95 del MINECOM y los dispuesto en los Arts. 1, 3, 4, 5, 6, 7, 8, 9, 10, 14, 16, 17, 18, 22, y 24 de la Ley 18.287, sobre procedimiento ante los Juzgados de Policía Local;

SE DECLARA:

Se condena a , ya individualizado, por haber infringido el D.S. 225 del 9 de Noviembre de 1995 al haber cometido actos de maltrato y la muerte de la especie mamífero denominado Chungungo o Gato de Mar (Lutra Felina), al pago de una multa ascendente a 10 U.T.M., la que deberá ser cancelada dentro de plazo en Tesorería Municipal bajo estricto apercibimiento legal.

Notifíquese y en su oportunidad archívese.

Resolvió el Juez de Policía Local Titular don Francisco J. Sanfuentes Gazteiu;

MARGARITA TORRES CISTERNAS , SECRETARIA

References

Aguilera, M. A., Aburto, J. A., Bravo, L., Broitman, B. R., García, R. A., Gaymer, C. F., Gelcich, S., López, B. A., Montecino, V., Pauchard, A., Ramos, M., Rutllant, J. A., Sáez, C. A., Valdivia, N., & Thiel, M. (2019). Chapter 9 – Chile: Environmental status and future perspectives. In C. Sheppard (Ed.), *World seas: An environmental evaluation* (2nd ed., pp. 673–702). Academic Press. https://doi.org/10.1016/B978-0-12-805068-2.00046-2

Alfaro-Shigueto, J., Valqui, J., & Mangel, J. C. (2011). Nuevo registro de la nutria marina Lontra felina (MOLINA, 1782) al norte de su distribución actual [New record of the marine otter Lontra felina (Molina, 1782) north of its current distribution]. *Ecología Aplicada, 10*(2), 87–91.

Barragán, J. M., & Lazo, Ó. (2018). Policy progress on ICZM in Peru. *Ocean & Coastal Management, 157*, 203–216.

Brownell, R. L. (1978). Ecology and conservation of the marine otter, Lutra felina. pp: 104–106. In N. Duplaix (Ed.), *Otters: Proceedings of the Firts working meeting of the otter specialist group*. WWF/SSC/IUCN. Paramaribo, Suriname 27–29 March 1977 (158p). Internacional Union for Conservation of Nature and Natural Resources, Publication new series.

Castilla, J. C. (1981). La nutria de mar chilena, especie en extinción. *Creces, 2*(1), 31–34.

Cursach, J. A., Rau, J. R., Ther, F., & Vilugrón, J. (2012). Sinantropía y conservación marina: El caso del chungungo Lontra felina en el sur de Chile. *Revista de Biología Marina y Oceanografía, 47*, 593–597.

Defensoría del Pueblo. (2022). *Derrame de Petróleo en la refinería La Pampilla*. Avalaible: https://www.defensoria.gob.pe/wp-content/uploads/2022/04/Informe-Derrame-Ventanilla.pdf

Duplaix, N., & Savage, M. (2018). *The global otter conservation strategy*. IUCN/SSC Otter Specialist Group.

ENAP Refinerías S.A. (2019). *Memoría Anual ERSA 2019* (126p). Empresa Nacional del Petrólero ENAP.

Fitter, R. (1973). Convention to control the rare animal trade. *Oryx, 12*(2), 188–194.

Foster-Turley, P., Macdonald, S. M., & Mason, C. F. (Eds.). (1990). *Otters: An action plan for their conservation*. IUCN.

Garay, J. (2003). *Marco Jurídico de la Reserva Nacional de Paracas y la Pesca Responsable*. Avalaible: https://tesis.pucp.edu.pe/repositorio/bitstream/handle/20.500.12404/1235/GARAY_RODRIGUEZ_JOHANA_JURIDICO_PARACAS.pdf?sequence=1&isAllowed=y

Gillete, R. (1973). Endangered species: Moving toward a cease-fire. *Science, 179*(4078), 1107–1109.

Glade, A. A. (1988). *Libro rojo de los vertebrados terrestres de Chile. Actas simposio "Estado de conservación de la fauna de vertebrados terrestres de Chile"*. Organizado por la Corporación Nacional Forestal entre el 21 y 24 de abril de 1987 en Santiago, Chile, 65p.

Godfrey, M. H., Roberts, D. L., & Godley, B. J. (2008). Taking it as red: An introduction to the theme section on the IUCN red list of threatened species. *Endangered Species Research, 6*, 109–111. https://doi.org/10.3354/esr00168

Instituto de Fomento Pesquero IFOP. (2016, Septiembre). *INFORME FINAL "Determinación de los impactos en los recursos hidrobiológicos y en los ecosistemas marinos presentes en el área de influencia del derrame de hidrocarburo de Bahía Quintero, V Región"*. SUBSECRETARÍA DE PESCA Y ACUICULTURA IFOP , 711 p.

Iriarte, J., & Jaksic, F. (1986). The fur trade in Chile: An overview of seventy-five years of export data (1910–1984). *Biological Conservation, 38*, 243–253.

IUCN. (1964, April). *Preliminary list of rare mammals including those believed to be rare but concerning which detailed information is still lacking*. Compiled by the IUCN Operations Intelligence Centre, Morges. Published as a Special Supplement to IUCN Bulletin (new series) No. 11.

IUCN. (2012). *Categorías y Criterios de la Lista Roja de la UICN: versión 3.1, segunda edición*. Gland, d, Suiza y Cambridge, Reino Unido: UICN. Vi + 34pp.

IUCN. (2022). *Statutes of 5 October 1948, 21evise don 22 October 1996, and last amended on 10 September 2021 (including Rules of Procedure of the World Conservation Congress, last amended on 10 September 2021) and Regulations 21evise don 22 October 1996 and last amended on 10 September 2021.* IUCN. Gland. https://doi.org/10.2305/IUCN.CH.2021.SR.01.es

Jefferson, T. A., Webber, M. A., & Pitman, R. L. (2015). *Marine mammals of the world a comprehensive guide to their identification* (pp. 535–546). Academic Press.

Joslin, P., & Mayanka, D. (1968, March). *Endangered mammals of the world, reporto on status and action treatment* (IUCN Publications New Series. Supplementary Papers No 13). Survival Service Commission.

Kaplan, M. (1973, February 22). An ilegal fur ring broken up here by U.S. *The New York Times.*

Kruuk, H. (2006). *Otters ecology, behaviour and conservation.* Oxford University Press.

Larivière, S. (1998). Lontra felina. *Mammalian Species, 575*, 1–9.

Mangel, J. C., Whitty, T., Medina-Vogel, G., Alfaro-Shigueto, J., Cáceres, C., & Godley, B. J. (2010). Latitudinal variation in diet and patterns of human interaction in the marine otter. *Marine Mammal Science, 27*(2), E14–E25. https://doi.org/10.1111/j.1748-7692.2010.00414.x

Mangel, J., Alfaro-Shigueto, J., Verdi, R., & Rheingantz, M. L. (2022). *Lontra felina. The IUCN red list of threatened species 2022* (e.T12303A215395045). https://doi.org/10.2305/IUCN.UK.2022-2.RLTS.T12303A215395045.en

Medina-Vogel, G. (2010). Ecología de enfermedades infecciosas emergentes y conservación de especies silvestres. *Archivos de Medicina Veterinaria, 42*, 11–24. https://doi.org/10.4067/S0301-732X2010000100003

Medina-Vogel, G., Boher, F., Flores, G., Santibañez, A., & Soto-Azat, C. (2007). Spacing behavior of marine otters (Lontra felina) in relation to land refuges and fishery waste in Central Chile. *Journal of Mammalogy, 88*(2), 487–494. https://doi.org/10.1644/06-MAMM-A081R1.1

Olavaria, C., ALvarez R., Correa R., Badilla M., & Carrasco H. (2007). Impactos reales y potenciales de las actividades antropogénicas sobre mamíferos marinos en Chile: Nutrias. En: *CPPS/PNUMA*. 2007. Memorias del Taller de Trabajo sobre el impacto de las actividades antropogénicas en Mamíferos Marinos en el Pacífico Sudeste. Bogotá, Colombia, 28 al 29 de noviembre de 2006. Guayaquil, Ecuador. 98p.

Oryx. (1973). Notes and news. *Oryx, 12*(1), 3–14.

Ostfeld, R. S., Ebensperger, L., Klosterman, L. L., & Castilla, J. C. (1989). Foraging, activity budget, and social behavior of the south American marine otter Lutra felina (Molina 1782). *National Geographic Research, 5*, 422–438.

Ponce, C. (1973). Informe Nacional sobre Fauna en el Perú. En: *Instituto Interamericano de Ciencias Agrícolas da OEA Programa Cooperativo para o Desenvolvimento do Trópico Americano IICA-TROPICOS, Ministério da Agricultura.* Simposio Internacional sobre Fauna Silvestre e Pesca Fluvial e Lacustre Amazónica (Novembro 26–Dezembro 1, 1973). Manaus, Amazonas, Brasil. Informes de Conferencias, Cursos y Reuniones N° 26. Volume I.

Pulido, V. (1991). *El Libro Rojo de la Fauna Silvestre del Perú* (220p). Editorial Maijosa.

Servicio Agrícola y Ganadero. (2015). *Legislación, La Ley de Caza y su Reglamento.* Subdepartamento de Vida Silvestre, División de Protección de los Recursos Naturales Renovables, SAG. Ministerio de Agricultura.

Subsecretaría de Pesca y Acuicultura. (2018). *Estandarización metodológica para desarrollo de líneas base y seguimientos ambientales de mamíferos marinos.* Guía de Buenas Prácticas y Acciones de Mitigación para el seguimiento ambiental de mamíferos marinos para proyectos que se someten al Sistema de Evaluación de Impacto Ambiental en Chile. Avalaible: https://www.subpesca.cl/fipa/613/w3-article-101113.html

Valqui, J. (2012). The marine otter Lontra felina (Molina, 1782): A review of its present status and implications for future conservation. *Mammalian Biology, 77*, 75–83. https://doi.org/10.1016/j.mambio.2011.08.004

Chapter 8
Marine Otter (*Lontra felina*, Molina 1782) Conservation in Chile

Gonzalo Medina-Vogel, Carlos Calvo-Mac, Gabriela Molina-Maldonado, and Nicole Delgado-Parada

Abstract The marine otter or chungungo (*Lontra felina*) is one of the least known carnivore species in Chile, and the available information is scarce and dispersed in time. Few studies have been conducted in Chile before and after the revision of the Natural History of the species in 1998. Therefore, we re-examine the knowledge of the species with current evidence on the marine otter in Chile and the main threats described for its conservation. We observed an irregular pattern in its distribution associated with the size and distance of rocky patches on the coast along with human influence. Also, the southern limit of the distribution remains unclear, with scarce current evidence of the marine otter presence south of Chiloe Island. It has been observed that the marine otter modifies its diet according to the availability of prey, so this would be a limiting environmental variable for its population density and distribution. Threats to the conservation of the species have been identified related to the alteration of its habitat, including incidental mortality from fishing activities, persecution by domestic dogs, and the potential for disease transmission. There is a great need for studies focused on ecology and human dimensions of the marine otter, to carry out adequate conservation of the species.

Keywords Marine otter · Review · Human disturbance · Conservation

G. Medina-Vogel (✉)
Centro de Investigación para la Sustentabilidad, Facultad Ciencias de la Vida, Universidad Andrés Bello, Santiago, Chile

IUCN Species Survival Commission, Otter Specialist Group, Gland, Switzerland
e-mail: gmedina@unab.cl

C. Calvo-Mac
Centro de Investigación para la Sustentabilidad, Facultad Ciencias de la Vida, Universidad Andrés Bello, Santiago, Chile
e-mail: carlos.calvo@upch.pe

G. Molina-Maldonado · N. Delgado-Parada
Facultad de Ciencias Biológicas, Pontificia Universidad Católica de Chile, Santiago, Chile
e-mail: gvmolina@uc.cl; nadelgado@uc.cl

L. Ayala et al. (eds.), *Marine Otter Conservation*,
https://doi.org/10.1007/978-3-031-53931-2_8

8.1 Introduction

The marine otter or chungungo (*Lontra felina*; Molina, 1782) is the smallest of the two otter's species inhabiting Chile, and the only exclusively marine. The other is the southern river otter (*Lontra provocax*; Thomas 1908). Marine otter is one of the least known carnivore species in Chile, with the available information being scarce and scattered in time and geographic distribution. Few studies have been conducted in Chile before and after the first most complete review of the species Natural History, made by Lariviere (1998). This was based on historical references and both peer-reviewed papers and gray literature (e.g., unpublished, proceedings, and technical reports), so has to be interpreted with caution. The marine otter was first described in Chile by the Jesuit priest Juan Ignacio Molina (1782). Subsequent publications of the species are scarce. In 1835, during his five-year journey on HMS Beagle, Charles Darwin recorded marine otters at the Chonos Archipelago, Chile (Darwin, 1889). Gay (1874) carried out descriptive studies on the species (Gay, 1874). According to Housse (1953), marine otter distributed along all the Chilean seashore, including observation of the species along the docks of the seaport of Valparaiso city (33°01′S), between Algarrobo and Tunquén (33°19′S). Also observed south of Tomé (36°38′S), Tirúa (38°20′S), in the Ancud Gulf (41°51′S), Llancahé channel (42°03′S), Guaitecas and Chonos Archipelago (43°46′S to 46°36′S) (Housse, 1953). It was only at the end of the 1970s beginning of the 1980s that marine otter began to be studied in Chile (Castilla & Bahamondes, 1979; Sielfeld, 1983; Cabello, 1985). Here we re-examine the knowledge of the species, reviewing data from scientific articles and conference proceedings produced since 1980 of current evidence about the marine otter in Chile and main described threats for its conservation. Furthermore, we describe the species current distribution according to facts of the species being observed or studied. We were able to join documents that contained information about the species observations in the wild, samples analysis and evidence obtained from scientific practices.

This review is organized in three sections. The first is an overview of the marine otter current distribution; the second is about ecological data and describes threats; and the third discusses current knowledge gaps, research need and provides conservation guidelines.

8.2 Marine Otter Distribution and Ecology in Chile

Using the known field methods of direct observation of an otter in the water or on the ground, and the registration of otter field signs (scats, footprints), the first significant survey of the species in Chile was done by Walter Sielfeld along seashores of several islands in Magallanes, Southern Chile in the 1980s (Sielfeld, 1989, 1990a, b). Thereafter, the first continental survey was done by Medina-Vogel et al. (2008), between 2004 and 2006 authors studied the occurrence of marine otters along

~2000 km of continental seashore in Chile (Medina-Vogel et al., 2008). Marine otter lives along the Pacific Ocean from the Northern limit of Perú in the artisanal port of Huanchaco (8°04′S) (Alfaro-Shigueto et al., 2011) and in Salaverry (8°13′S) (Santillán & Caro, 2007) to Seno Monorayo (55°16′S) Cape Horn (Sielfeld, 1990a, b), inhabiting the rocky seashores exposed to winds and waves (Castilla & Bahamondes, 1979; Ostfeld et al., 1989). Furthermore, its activities are confined to an area not wider than 50 m on land and 150 m offshore (Castilla & Bahamondes, 1979; Ostfeld et al., 1989; Medina-Vogel et al., 2006, 2007). Indeed, the species register a patchy distribution from Perú to Cape Horn (Chile), associated to a variety of habitat physical characteristics and prey availability (Medina-Vogel et al., 2008). Within their home range, their most important limited resource is safe shelter inside rock cracks and caves close to food resources. Its home range is small (<4.5 km long) and is affected by human use of the marine littoral zone (Medina-Vogel et al., 2007; Mangel et al., 2011). In fact, this otter must coexist with humans by inhabiting the marine littoral zone and exploiting the same resources from the intertidal, subtidal and terrestrial seashore that humans exploit for food, commerce and housing (Moreno et al., 1984; Ostfeld et al., 1989; Moreno, 2001; Medina-Vogel et al., 2004, 2007). Therefore, the abundance of marine otters in one patch of rocky seashore, is the result of the habitat and surrounding matrix quality. In fact, in an anthropogenically fragmented landscape, patch area and isolation have been named as the two strongest landscape predictors of top terrestrial predator distribution and abundance (Crooks, 2002). Likewise, marine otter populations isolation is associated with human influence on the seashore and size of the rocky seashore patches, becoming more important the effect when the patches were smaller and farther apart from each other (Medina-Vogel et al., 2008). Indeed, rocky seashore patch networks containing rocky patches over 5.1 km long close to each other could contain home ranges of several individuals. Thus, rocky seashore patches separated between 9.2 km and 36.8 km do not show any significant difference in otter field signs occurrence. Instead, isolated (more than 40 km apart) patches smaller than 15 km long could therefore be considered sinks and might be dependent on high level of immigration for their continuation (Medina-Vogel et al., 2008). Supporting this observation is the fact that marine otters spend more than 80% of their daily time on land inside secure rock caves or cracks, where most of their time is spent resting (Medina-Vogel et al., 2006, 2007). Indeed, the marine otter is the smallest of its genus, and the most recent mammal adapted to a marine environment (Estes, 1989). Contrary to *Enhydra lutris*, which distribute from North America towards Asia and could spend all its life in the water, the marine otter spends most of its daily time out of the water resting inside caves or rock cracks (Medina-Vogel et al., 2006, 2007). The little time that marine otter consumes in the water may be due to an evolutionary adaptation or a behavioral response to an environment with abundant prey close to shore, where dives into the water are short and highly successful (Ostfeld et al., 1989; Medina-Vogel et al., 2004). If an animal spends a very limited amount of time in the water, then it does not have the opportunity to disperse by swimming along the shoreline to avoid a sandy beach environment that may be hostile due to dogs and human activity (Medina-Vogel et al., 2008). In fact, molecular studies suggest two geographic barriers to marine otter dispersion along continental Chile: (1) from Algarrobo (33°22′S) south toward Peninsula Tumbes (36°36′S) and (2) south from Arauco Bay (37°10′S) to Queule (39°23′S). Both areas correspond to the longest sandy beaches present in Chile with few small patches of rocky seashore where marine otter is nearly absent (Vianna et al., 2010). Indeed, Vianna et al. (2010) study showed a high dependence of marine otter on large rocky seashore patches to maintain genetic diversity. Geographically rocky seashore patches are geographically separated increasingly, from northern Chile to central-south Chile, and in this region the marine otter occupancy of rocky seashore is significantly influenced by the presence of humans, in fact marine otter occupancy on surveyed rocky seashore patch was significantly lower in those sections and in rocky patches associated with high levels of human influence of the seashore (Medina-Vogel et al., 2008). Also, the significantly higher marine otter occupancy was recorded only in the largest rocky seashore patches (~412 km long) (Medina-Vogel et al., 2008). As expected, a pattern between increasing distance to the nearest large rocky seashore patch and decreasing marine otter occupancy is not observed in those regions with high human influence (Medina-Vogel et al., 2008). Consequently, when analyzing patch size (length) against isolation, it becomes clear that the size and isolation of the rocky seashore network within an area are significantly associated with the occupancy incidence of marine otters. This relationship is altered by the human influence on the seashore (Medina-Vogel et al., 2008). Furthermore, the diet of marine otter become more piscivorous towards the north, where occupancy incidence of marine otters is higher (Medina-Vogel et al., 2008; Mangel et al., 2011). Thus, according to the previous information and data, a distribution map for marine otter in Chile should consider first the extension of the rocky seashore patch, second their distance to the contiguous ones, and third a confirmation of the presence of the species. Nevertheless, marine otter confirmed observation reports are from Los Molles (32°13′S) and Yerba Buena (29°32′S) (Castilla & Bahamondes, 1979). Conspicuous groups are reported in Mejillones (23°04′S), Constitución (23°54′S), Pan de Azucar (24°40′S), Soldado Bay (26°10′S), Isla Choros (29°15′S), Arrayán (29°43′S), Panul (30°03′S), Las Tacas (30°05′S), Palo Colorado (32°01′S), Puquén (32°11′S), Papudo (32°26′S), Cachagua (32°35′S), Quintay (33°11′S), Hualpén (36°46′S), Tumbes (36°42′S), San Vicente (36°43′S), Chome (36° 46′S), Colcura (37°07′S), coast of Valdivia (39°39′S), Pucatrihue (40°10′S), Bahía Mansa (40°34′S), Punhihuil (41°55′S), Chiloé Island (42°10′S), Quilán Island (43°23′), and Guafo Island (43°35′) (Rozzi & Torres-Mura, 1990; Villegas et al., 2007; Córdova et al., 2009; Badilla & George-Nascimento, 2009; Vianna et al., 2010; Cursach et al., 2012; Poblete et al., 2019; Gutiérrez et al., 2019; Calvo-Mac et al., 2020). Between Algarrobo (33°22′S) and Peninsula Tumbes (36°36′S) and between Arauco Bay (37°10′S) and coast of Valdivia (39°23′S) no marine otters were reported (Fig. 8.1) (Medina-Vogel et al., 2008; Vianna et al., 2010). Very little is known about the populations south of Guafo Island and Chiloé Island (43°35′S).

Between January and December 1982, Sielfeld studied 11 sites with a total of 67 marine otter dens, seven in the Mornington Island (49°42′S), six in the Contreras Island (52°00′S), two in the Bauclork Island (53°11′S), seven in the Carlo II Island

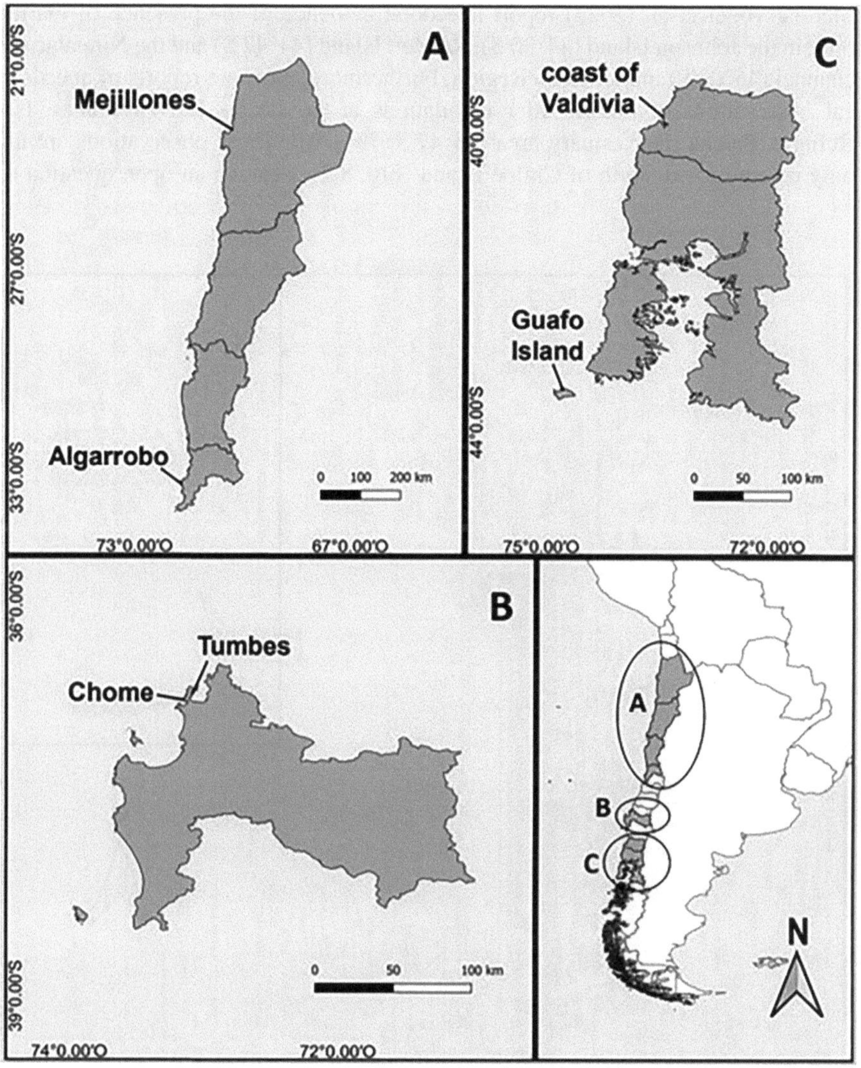

Fig. 8.1 Sighting areas of *Lontra felina* reported in Chile. (**a**) Sighting area located between the Antofagasta region (Mejillones) and the Valparaíso region (Algarrobo). (**b**) Sighting area located in the Biobio region with reports only between Chome creek and the Tumbes peninsula. (**c**) Sighting area located between Los Rios region (coast of Valdivia) and Los Lagos region (Guafo Island)

(53°40′S), five in the Skyring (54°23′S), four in the Capitan Aracena Island (54°10′S), four in the Baskot Island, eleven in the Stewart Island (54°52′S), six in the Brocknock channel (54°39′S), eight in the Hosto Island (Seno Monorayo (55°16′S) and seven dens in the Grevy island (56°S) (Cape Horn) (Sielfeld, 1990a, b). There are no new records for marine otters in Magallanes. Recently

Medina-Vogel et al. (2023) report anecdotal evidence of the presence of marine otter in the Johnson Island (44°20′S), Rowlett Island (44°47′S) and the Ninualac sea channel (45°02′S) in the Aysén Region. Furthermore, there are reports of anecdotical observations of marine otter populations at the Islotes Las Hermanas, Isla Refugio, Palena river estuary area (43°47′S, 73°1′W). These observations are the only recent record south of Chiloé Island (Fig. 8.2). Leaving an open question of

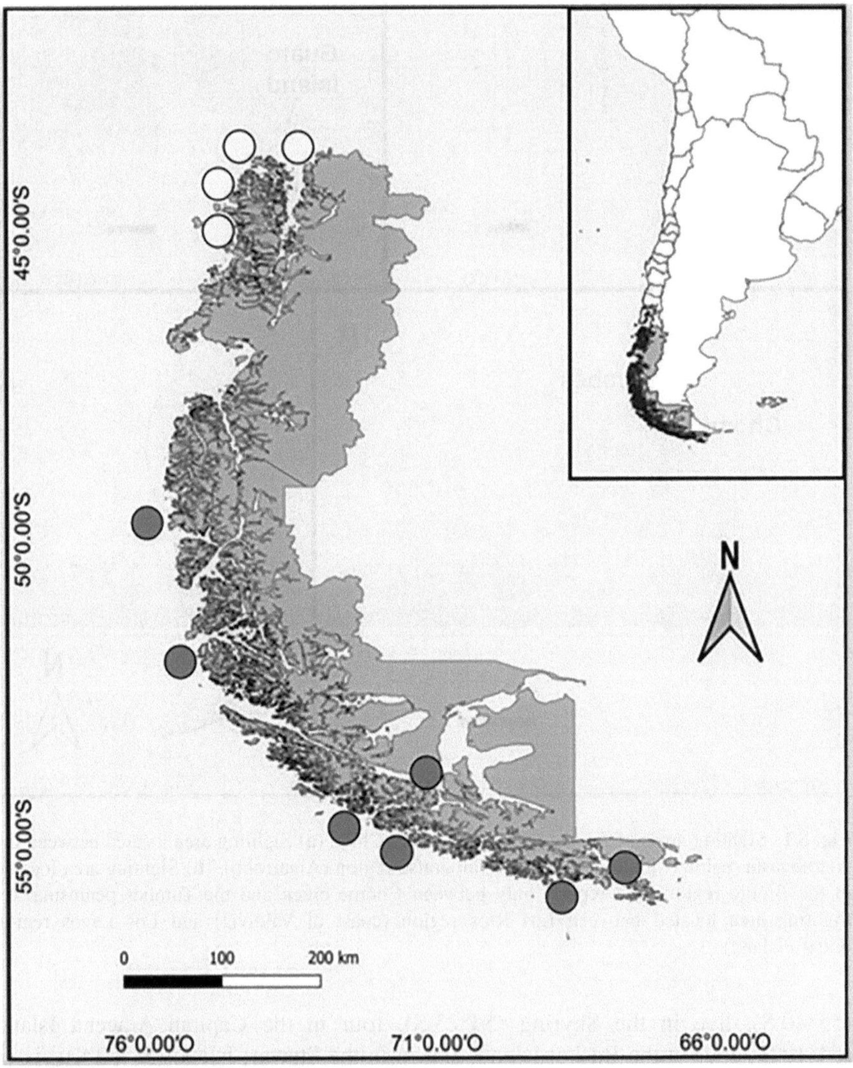

Fig. 8.2 Sighting reports of *Lontra felina* in the region of Aisén and Magallanes (Chile). The sightings are represented as circles, where the gray circles correspond to previously reported sightings, while the white circles correspond to recently reported sightings

how the marine otter populations are distributed between the west coast of Chiloé Island and the Palena river estuary. The patchy distribution of the marine otter, although provides an opportunity to investigate the relationships among patch size, isolation and human impact, this patchy distribution and increased population isolation, is thought to be forcing the species to decline and currently it is listed as 'Endangered' by the International Union for the Conservation of Nature and Natural Resources (IUCN) (Mangel et al., 2022) and "Endangered" by the Chilean National Species Inventory (Ministerio del Medio Ambiente, 2021).

8.3 Marine Otter Conservation in Chile

In Chile as well as in Perú, the conservation threats to marine otter are still poorly understood and, despite being one of the targets species in any Environmental Evaluation Assessment for any seashore investment project, the small number of studies along its distribution, and lack of comparative studies, generally precluded further analysis, and end in assessments comparing studies with large differences in sample size and observation time (e.g., Córdova et al., 2009). Exceptions are the studies of Medina-Vogel et al. (2007, 2008), that were species-oriented. The first investigated in a section of the seashore of Quintay (33°11′S), north/center of Chile, daily activity patterns, social and spatial of six radio-tracked marine otters (three adult males, three adult females). Contrary of what most of the previous studies suggest, the results indicated marine otter as a more nocturnal species (~60%) with individuals frequently resting during night and day inside burrows, rock cracks, or underneath rocks (~80%). The long-time marine otter spend out of the water was already recorded before by Medina-Vogel et al. (2006). Marine otter home ranges and core areas followed the seashore contours. Indeed, recorded home range were between 1373 m to 4134 m long and less than 130 m wide (102 m offshore and 30 m on land), with no differences between sexes. Core areas varied from 49 m to 495 m long with no significant differences between sexes. Core areas were associated to caves or cavities between rocks or rock cracks used as dens or resting places on land (Medina-Vogel et al., 2007). Core areas averaged 8% of home ranges of males and females. Marine otters can move among 55–151% of their home range extension in 24 h. In fact, female marine otters exhibited intrasexual territoriality, but males showed no territoriality. Marine otter with or without transmitters were observed simultaneously sharing a group of rock cavities in an area not larger than 100 m^2. However, marine otters are solitary, there are no records of family groups, cooperative feeding, or groups of otters moving or resting together in the same den or feeding patches at sea (Medina-Vogel et al., 2007).

8.3.1 Threats for Marine Otter Conservation

Between 2009 and 2023, 130 otters were registered and handled by SERNAPESCA, 34 alive, 96 dead. The number of otters recorded dead or alive has been increasing since 2009 (Fig. 8.3). Unfortunately, most individuals were not classified by their sex and body characteristics (length, weight). The larger proportion of marine otters were registered from the central to the north of Chile. Most of the cause of death was registered as indeterminate, alive 34 (25%) from which only 19 (55%) were reintroduced, so 15 (45%) otters that were found alive died during the rehabilitation process. A total of 25 (18%) were registered as drowned, 20 (15%) as potentially being attacked or predated by domestic dogs, 6 (4%) run over or crushed, and 1 (1%) intoxicated by hydrocarbons. An increased tendency of otter death by indeterminate cause were registered towards 2023 (Fig. 8.3). Most of the animals were found in the proximities of crowded cities and towns (98%) (Fig. 8.3). And most of them on sandy beaches 75 (58%) and 35 (27%) inside a human-made infrastructure, form inside the motors of a vehicle, inside buildings, to inside the refrigeration system of thermoelectric centrale (Correa & Pizarro, 2023). This last increase of marine otter death coincides with the High Pathogenic Avian Influenza A (H5N1) outbreak

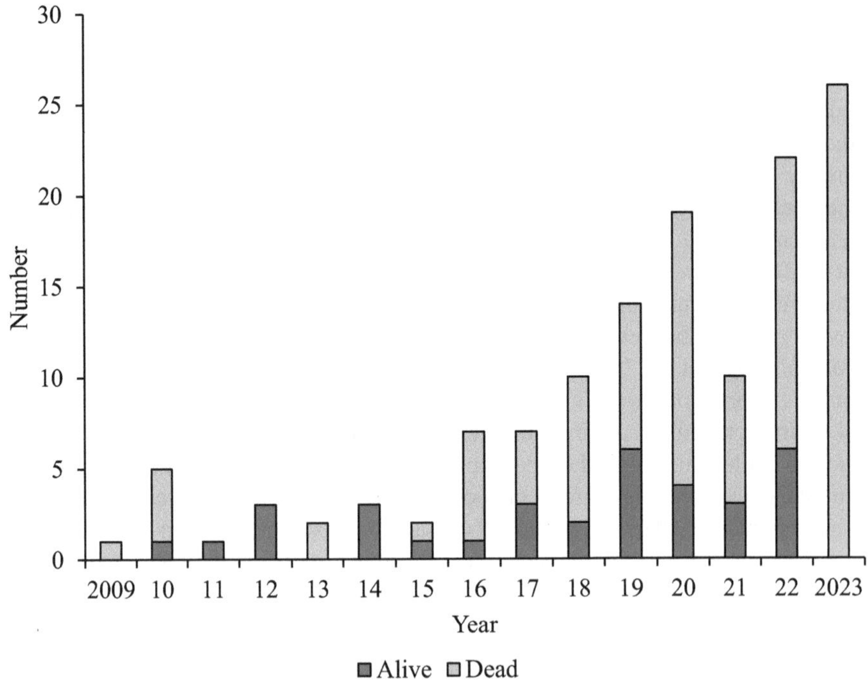

Fig. 8.3 Annual increase between January 13th, 2009 and May 17th, 2023 of marine otters (*Lontra felina*) found dead or alive in Chile registered by SERNAPESCA

along the south Pacific region. In fact, all otters registered between January 17th and May 17th 2023 were found dead.

8.3.1.1 Fisheries Interactions

Marine otter modifies its diet according to prey availability and so its diet differs along its distribution (Medina-Vogel et al., 2004; Córdova et al., 2009; Alfaro-Shigueto et al., 2011). Between Magallanes (48°50′S), Chiloe Island (42°10′S) and Pan de Azucar (24°40′S) crustaceans are the most consumed preys, followed by fish and, to a lower extent, mollusks (Córdova et al., 2009). In fact, there is a latitudinal pattern regarding prey consumption. Maximum values (Shannon-Wiener, H′) are found in the central-south latitudes and minimum values toward the south and north (Córdova et al., 2009). Marine otter had a more crustacean-based diet in the central-south latitudes and more fish-based diet towards the southern (Magallanes) and North distribution (Mangel et al., 2011). Nevertheless, comparative studies show that marine otter predate mostly on crustaceans, followed by fish and, to a lower extent, mollusks (Córdova et al., 2009). Thus, marine otter is an opportunistic predator. Feeding mainly on the most abundant subtidal prey species, responding to seasonal and site variations in prey availability (Medina-Vogel et al., 2004; Villegas et al., 2007). Thus, for marine otter total time spending searching for food is a limitation environmental variable for its population density and distribution. It is important to consider that marine otter must deal with heat loss imposed by the cold environment of the south Pacific Ocean. Marine otter being the smallest of its genus (*Lontra provocax*, *Lontra longuicaudis*, *Lontra canadensis*), it might control body heat loss by reducing total time in the water, focusing the time on the water mostly for feeding (Medina, 1993; Medina-Vogel et al., 2007). Indeed, marine otter spent less than 30% of its daily activity feeding (Medina-Vogel et al., 2007), and its diet contain mostly seabed-dwelling species, small invertebrates, highly available, easy to capture and with low handling time to eat (Medina-Vogel et al., 2004). Moreover, there is no evidence of competition between *L. provocax* and *L. felina*, in fact Ebensperger and Botto-Mahan (1997) suggested that both species segregate habitats to avoid interspecific competition for food, so how these two species share the area between south Chiloé Island and the Palena river estuary, where both species have been describing, remain a mystery (Ebensperger & Botto-Mahan, 1997). However, the marine otter highly dependent on diet to intertidal, subtidal seabed-dwelling species makes humans the most powerful competitor for marine otter. Although marine otter preys on resources exploited by artisanal fishing, without damaging their fishing gears. Indeed, a study done in Huenteyao (40°28′S; 73°43′W), report that 66.7% of the fishermen said that otters do not affect their fishing, while 30.0% said the opposite; holding as cause the consumption by marine otter on the same resources exploited by them (e.g., *Concholepas concholepas*) (Córdova & Rau, 2016). Furthermore, there is a synanthropic behavior of marine otter along its geographic distribution. The marine otter used fishing ports infrastructure for shelter and food. It forages on the remains of fish discarded by fishermen which affect the otters' spatial and social behavior (Medina-Vogel et al., 2007; Cursach et al., 2012). Correa and Pizarro (2023) recently reported three cases of otter death as a result of being caught and drowned by fishing and crab gears, three cases of otter death presumably as a result of being run over by boats. Furthermore, when fishery waste from fishing landings is available, marine otters have been observed together feeding on waste and sometimes being persecuted by street dogs (Medina-Vogel et al., 2007). Moreover, and possibly as part of the effect of human waste on the otter diet, recently a diverse type of microplastics were found in marine otter scats collected in Punta Corrientes (12°57′S), located in the central coast of Peru (Santillán et al., 2020).

8.3.1.2 Interaction with Domestic Dogs and Cats

Identified threats to the conservation of this species in synanthropy include persecution by domestic dogs and the potential of disease transmission. In fact, *T. gondii* exposure in one of 19 marine otters along the northern and central coast of Chile was registered (Calvo-Mac et al., 2020), indicating a possible low occurrence in seashore associated to arid environment. Furthermore, as has been registered in other otter species, marine otter can also be vulnerable to *Leptospira* spp. canine adenovirus, canine distemper virus, and feline influenza (Medina-Vogel, 2010; Barros et al., 2022). Correa and Pizarro (2023) reported four cases of otter wounded and possibly killed by domestic dogs. Interaction, like otters with dogs and cats in close proximity or using same waste feeding spots from human activity, has been observed in Chile and in Peru (Medina-Vogel et al., 2007; Pizarro-Neyra, 2008; Mangel et al., 2011).

8.3.1.3 Habitat Lost and Fragmentation

Marine Otter Habitat Medina-Vogel et al. (2008) assessed the species' ecological requirements and identified landscape limitations for the marine otter occupation and dispersal, in a study area of about ~2000 km long, length of sandy beaches, size and distance between rocky seashore patches, and human use of the seashore, were evaluated as the main habitat features responsible for the marine otter distribution and occupancy patterns in the continental seashore exposed to the Pacific Ocean. The occurrence of marine otters varied significantly depending on the distance to the nearest rocky seashore patch, reducing the probability of occupancy of a rocky seashore patch when the distance to the closed one is above 14 km apart. Indeed, rocky seashore patch networks containing rocky patches over 5.1 km long closer than 14 km apart to each other could contain home ranges of several marine otters. In fact, rocky seashore patches separated by <20 km and larger than 15 km long could therefore be considered good core habitat network for marine otter (Medina-Vogel et al., 2008). Furthermore, these authors described the highest marine otter occupancy in the largest rocky seashore patches (~412 km). And as expected, a pattern of decreasing marine otter occupancy associated to an increasing distance to the nearest large rocky seashore patch and greatest human occupation of the seashore. Nevertheless, marine otter preference for rocky seashore is inconsistent with what have been observed when comparing exposed with not exposed rocky seashore to sea weaves. Based on the presence of otter field signs (scats, dens, direct observation on land), marine otter is said to prefer wave-exposed habitats over more wave-protected areas of the coast and sandy beaches (Sielfeld, 1983, 1990a; Ebensperger & Castilla, 1992), and this greater use of wave-exposed habitats is suggested to reflect differences in the availability of prey taken by the otter (Ebensperger & Castilla, 1992; Medina-Vogel et al., 2004). In fact, marine otter diet concentrated on benthic dwelling preys, so dive time depends on depth and hunting success (Ostfeld et al., 1989; Medina, 1995; Medina-Vogel et al., 2004). Moreover, diving success is expected to increase with prey abundance, a measure of patch quality, and optimality considerations predict that diving time should increase with depth (a surrogate of travel time), decrease with energetic costs of diving. In fact, data provided from Choro Island (29°15′S, 71°32′W) shows that crustaceans were more abundant in the wave-exposed habitat, but fish were more abundant at the wave-protected site (Villegas et al., 2007). There, marine otters spent more time foraging in the wave-protected site compared with the wave-exposed habitat. Successful dives reached 26.9% in the wave-exposed habitats, and 38.2% in the wave-protected habitat. Foraging dives were 18% shorter in wave-exposed as compared with wave-protected habitat (Villegas et al., 2007). And as numerically, available prey did not differ significantly with habitat, Villegas et al. (2007) results were more consistent with the hypothesis that wave-exposed habitats represent a sub-optimal habitat to foraging marine otters. But majority of dives of marine otters recorded before at Curiñanco (39°43′S, 73°24′W) were between 25% and 44% successful, and 10% were the longest (55–64 s) and were ~ 50% saucerful. This is less of what was recorded at Chiloé Island by Ostfeld et al. (1989), so otters from Curiñanco and Choro Island spend more on feeding than at Chiloé Island, also otters at Choro Island spend similar time feeding that at Los Molles (32°14′S; 71°31 W) (Ostfeld et al., 1989). Furthermore, marine otter from Curiñanco do not select prey based on total energy content (Medina-Vogel et al., 2004). Also, contrary to what was observed in Choro Island, in Talcahuano (36°44′S, 73°11′W) otters recorded longest dives in those wave-exposed seashores than those wave-protected habitat, even otters were observed more frequently consuming their prays on the water instead of inland in the more wave-exposed site (Badilla & George-Nascimento, 2009). Therefore, dive length is likely to be related to water depth and wave-exposed habitat and not to prey quality and availability, consequently, if the dive length is affected mainly by water depth and condition, and the selection of the area where otters began dive is random, dive success and time spend feeding will differ markedly from areas with different water depth and environmental conditions (Medina, 1993; Villegas et al., 2007; Badilla & George-Nascimento, 2009). Indeed, marine otters' use of wave-exposed patches through northern and central Chile coastal areas probably reflects a low availability of suitable protected areas and greater human disturbance of more protected habitat (Villegas et al., 2007).

Habitat Loss and Fragmentation Habitat use by marine otter is selective and most probably respond to the changes caused by humans (Medina-Vogel et al., 2007, 2008; Cursach et al., 2012; Gutiérrez et al., 2019). Indeed, there is a strong negative association between marine otter occupancy and human presence. Furthermore, there is a pattern between increasing distance to the nearest large rocky seashore patch and decreasing marine otter occupancy when more human presence and activity is recorded. Consequently, when analyzing patch size (length) against isolation, it becomes clear that the size and isolation of the rocky seashore network within an area are significantly affected by the human presence and activity between rocky patches (Medina-Vogel et al., 2008). Indeed, when analyzing SERNAPESCA data, most of otter found dead or alive where on sandy beaches 75 (58%), and 35 (27%) inside a human-made infrastructure, inside the motor of a vehicle, or inside the refrigeration system of a thermoelectric central.

8.3.1.4 Population Decline

Inconsistence among population estimation studies, in Chile, by using minimum abundance indices, that is counting the maximum numbers of otters active at the same time in a defined area for one year by three observers (June 1999–June 2000), involving a total of 320 hours of observation along three separated localities near Curiñanco, the average estimated density was 3.8 otter/km over a study area of approximately 4 km long (four separated but nearby study sites of 1 km long) (Medina-Vogel et al., 2006). Very close to the result of Biffi and Williams (2017) in Perú using molecular tools over 18.4 km of seashore (five separated study sites). Furthermore, the single radiotracking study in central Chile done in Quintay demonstrate a high variation in otter/km as a response to food availability, from 0 to 6 otters/km with an average otter density of about 1.7 otters/km of seashore, over a study area of 6.6 km long (Medina-Vogel et al., 2007). Also, Medina (1993) calculated an average of 8.3 otters/km if the observation was over a 1 km of seashore, and 4.4 otters/km if the observation was over 3 km of seashore in the seashore of Curiñanco, Valdivia (Medina, 1993). Thus, there have been several intentions to determine marine otter population densities and abundance along its geographical range, even to count them and compare between local groups. However, the lack of sexual dimorphism, external natural marks, the species elusive behavior, daily (night and day) activity patterns, and spatial behavior (core area and home range overlap) are fundamental obstacles for accurate population estimates and underlined the strong variation of density values obtained in several studies done in past years (Sielfeld, 1992; Medina-Vogel et al., 2007, 2008; Valqui, 2012). Thus, results about estimating population densities (marine otters/km) by direct observation must be interpreted with caution. Furthermore, most studies assessing the marine otter population density only covered limited regional areas, did not consider the distance covered by observers, and so applied different methodologies to determine the total population, making impossible any comparison (Sielfeld & Castilla, 1999; Valqui,

2012). Nevertheless, Torres et al. (2002) report 19 censuses along Chile, with an average 1.5 otters/km, north of Chile (administrative regions I-III), 2.3 otters/km, central Chile (regions IV-V), 7.1 otters/km, south of Chile (X region), and 1.1 otters/km southern Chile (XII region) (Torres et al., 2002); ranging from 0.04 otter/km to 10 otters/km (Castilla, 1982; Cabello, 1985; Rozzi & Torres-Mura, 1990). Additionally, there are a lot of contradictions concerning reproduction. Lariviere (1998) described the mating season for marine otters to occur between December and January, and parturition to occur from January to March. Cabello (1985) in Chiloé Island observed mating from December to January, pregnancy March–June, parturition between May and August. Medina-Vogel et al. (2007) in Quintay recorded one lactating female between May and July. Gutiérrez et al. (2019) recorded family groups in summer and fall. Cabello (1985), Medina (1993), Valqui (2004) and Medina-Vogel et al. (2006) describe pups of family groups all year round. And it is unknown whether *L. felina* exhibits delayed implantation (Lariviere, 1998). Female marine otters are usually seen swimming with one to two cubs. This inconsistency in determining marine otter population size is a threat to its conservation as it makes very difficult to assess population trends along its distribution. Moreover, there is no data about population size and distribution south of Chiloe Island to Cape Horn, where the species is generally confused with Southern river otter and American mink (Medina-Vogel et al., 2023).

8.3.1.5 Population Isolation

de Ferran et al. (2022) investigated genome-wide historical trends in effective population size (N_e) among 11 otter species. Among the 13 species of otters, the most extreme declines were for the Southern river otter (*L. provocax*) and the marine otter, likely driven by habitat changes (e.g., extensive ice sheet coverage) during the Last Glacial Period that may have severely affected their population numbers, particularly given that *L. felina* has a linear distribution restricted to rocky shores along the western coastline of South America. In Chile, a small number of marine otter haplotypes are shared between neighboring regions following a pattern of isolation by distance and confirming its low dispersal ability (Vianna et al., 2010). In fact, there are no concatenated haplotype shared between North-Peru, Norte (Grande and Chico), Central, Central-South, and South (Vianna et al., 2010). Indeed, apparently, two geographic barriers seem to prevent marine otter dispersal and so gene flow: (i) from Algarrobo (33°22′S) south toward Peninsula Tumbes (36°36′S) and (ii) south from Arauco Bay (37°10′S) to Queule (39°23′S) (Vianna et al., 2010). Both areas correspond to the longest sandy beaches present in Chile with few small patches of rocky seashore where marine otter is absent (Medina-Vogel et al., 2008; Vianna et al., 2010). Furthermore, a high population structure is observed at a large but also at a more local scale along the Chilean distribution of the species. At the fine scale, the population structure was dependent on the patchy structure of rocky seashore, mainly the distance between them. In fact, at Pan de Azucar (26°08′S–26°18′S).

Vianna et al. (2010) described a low population structure compared with south study sites in La Serena (29°37′S–30°17′S) and Palo Colorado (32°01′S–32°30′S), these last two study areas are characterized by short distances (i.e., 1.64–4.51 km) between extended rocky seashore patches, equivalent to the species home range of 3.2–4.2 km of lineal coast (Medina-Vogel et al., 2008). Indeed, Vianna et al. (2010) study showed a high dependence of marine otter on large rocky seashore patches to maintain genetic diversity along Chilean coastline.

8.4 Prospects and Challenges

8.4.1 Future Research

It is important to note that studies without tagged marine otters do not consider the time marine otters are out of view, so they are likely to be biased in favor of foraging. Also, studies about marine otter behavior are done during daylight, representing <10% of their daily movements, interspecific interactions, hunting behavior, resting, breeding and social behavior (Medina-Vogel et al., 2007), so those studies should be interpreted with caution. Nevertheless, several daylight observational studies stated that the marine otter displays activity peaks at specific hours during the day (Ostfeld et al., 1989; Medina, 1995; Badilla & George-Nascimento, 2009). During long-distance swimming between rocky patches sandy beaches are eventually used for resting (Rozzi & Torres-Mura, 1990; Ebensperger & Castilla, 1992). Otter nests are built and rebuilt repeatedly after hightide washed them away (Valqui, 2012). If captured prey are larger than the otters' head, they are carried in the mouth and may be put on belly while swimming on their back and brought to shore (Medina, 1995). Prey is then consumed with the help of the forepaws (Medina, 1995; Valqui, 2012). Thus, scientific information about the marine otter in Chile is scarce, geographically dispersed, with almost no information between Chiloé Island and Magallanes (Medina-Vogel et al., 2023). Furthermore, studies have concentrated on diet and behavior during daylight, so basic knowledge is needed to understand the current state of marine otter populations in Chile, especially south of Chiloé Island, and carry out conservation actions accordingly. The more pressing issue is establishing an occurrence map of the species in Chile, and how this occurrence differs according to the rocky seashore patches length and spatial distribution, island size and geographical distribution (south of Chiloé Island), and human occupation (residential, mining, energy generation industry, aquaculture, fishing, seaweed forests exploitation). This can be done by setting up a collaborative network to collect new data nationwide, for example using online questionaries and web resources in order to incorporate non-expert "citizen scientists" to make substantial and active contributions to species mapping and find threats to its conservation (e.g., Medina-Vogel et al., 2023). Furthermore, monitoring programs should be used by combining methodologies to ascertain marine otter presence, such as detection of scats, hair trapping, genetic species identification of scats, and camera-trapping campaigns. An example of successful large extension effort to evaluate marine otter distribution and patterns of occurrence by scats and live detection is the survey done by Medina-Vogel et al. (2008), methodology later supported by genetic identification in scats and tissue samples (Vianna et al., 2010). Another example to assess population density and distribution by genetic identification are those done in Perú using scats by Valqui et al. (2010) and Biffi and Williams (2017) (Valqui et al., 2010; Biffi & Williams, 2017). Furthermore, there is a need for standardization of visual estimation census. In this regards, Medina-Vogel et al. (2006) methodology proved to be trustworthy, as come to similar conclusions with the later study applying radiotracking and animal tagging of Medina-Vogel et al. (2007), marine otters spend more than 80% of their time inland. Once a distribution baseline is established, scats and hair trapping can be employed to assess dispersion and/or isolation between marine otter population, depending on the rocky seashore patches distribution and human use of the seashores (Vianna et al., 2010). Into this respect, further studies to assess how habitat modified by humans and artificial habitats (shipwrecks, docks, etc.) are used by marine otter as steppingstones between rocky seashore patches widely separated from each other are needed.

Future research on the marine otter in Chile should also aim to determine physiological limitations (osmoregulation) of the species to the sea conditions and prey availability, local abundance, sex ratio, dispersal rate, and expand incipient knowledge on the use of time and space by the species in a 24 h cycle. Also, Vianna et al. (2010) recommended further investigations on male dispersal patterns using biparental or paternal markers to better understand dispersal and the effective role of habitat structure. Additionally, there is increasing importance of further studies on disease infection from domestic animals and alien introduced American mink (*Neogale vison*), pathogen pollution (Calvo-Mac et al., 2020; Barros et al., 2022), and the ecology of pollution in general, e.g., how biomagnification affects marine otter as a top predator of south Pacific seashore.

8.4.2 Conservation Guidelines

At the moment there is no Recovering, Conservation and Management National Plan (RECOGE) specific for the marine otter in Chile. The species is classified as Endangered by the Chilean Ministry of Environment (https://mma.gob.cl/wp-content/uploads/2021/10/Acuerdo_N27_2021.pdf), Appendix I of CITES and Appendix I of the Convention for Migratory Species (https://www.cms.int/es/species/lontra-felina). Marine otter shared a small number of haplotypes between neighbor regions, following a pattern of isolation by distance (Vianna et al., 2010). In fact, there is no concatenated haplotype shared between Central-South and South

Chile, apparently because of two geographic barriers or extensive sandy beaches associated to high human occupation (Vianna et al., 2010), one from Algarrobo (33°22′S) south toward Peninsula Tumbes (36°36′S) and the second south from Arauco Bay (37°10′S) to Queule (39°23′S). And there is no data south of Chiloé Island. Furthermore, there is the urgent need to study the effect of salmon farming on marine otter ecology and conservation south of Chiloé Island. Marine otters are relatively easy to observe during daylight and from uplands. The species possesses a special status as flagship and umbrella species, which should be exploited to support educational campaigns and ecotourism along its distribution range. Sensitization of local people (especially fishers and salmon farming) towards marine otter might lead to conservation programs potentially contributing to the conservation of marine ecosystems in Chile. It is strongly recommended to coordinate efforts between research, environmental working groups, and stakeholders (Valqui, 2012). In Chile, the natural environments protected and managed by the state for their conservation are categorized according to the National System of State Protected Wildlife Areas (SNASPE), as: Parque Nacional, Parque Marino, Reserva Nacional, Reserva Marina, Monumento Natural, Santuario de la Naturaleza, Area Marina Costera de Multiples Usos y Espacio Marino Costero de los Pueblos Originarios (ECMPO). Each category must meet specific objectives such as protecting specific fauna. According to the records of marine otter *L. felina* in previous studies carried out in recent years and presented in this chapter, only eight protected areas reported the occurrence of marine otters (Fig. 8.4). Furthermore, not all protected areas have the marine otter as a species of conservation concern, and any of those areas include the marine area associated to them, they only protect land above sea level.

The hinder consequences of fragmenting forces acting upon marine otter suggested that conservation based on scattered protected areas will not work, instead management should consider the need of establishing reserve networks including the protection of rocky seashore patches that would allow migration (gene flow) and secure the species survival. Domestic animals should be controlled within and between these areas (Medina-Vogel et al., 2008; Mangel et al., 2011; Valqui, 2012). Considering political and logistical implications, this idea seems very difficult to put into practice (Valqui, 2012). However, marine otter as flagship and umbrella species could help in the development of a marine protected area network along the Pacific Seashore which will also help in the conservation of commercial invertebrates and fish of interest to the local fishers. Furthermore, it is urgent to run a marine otter occurrence/absence study south of Chiloé to Cape Horne, scat collection and population genetic studies. That region is currently suffering from the impacts of the salmon industry, and there is no reliable information of the conservation status of marine otter in all that region Medina-Vogel et al. (2023).

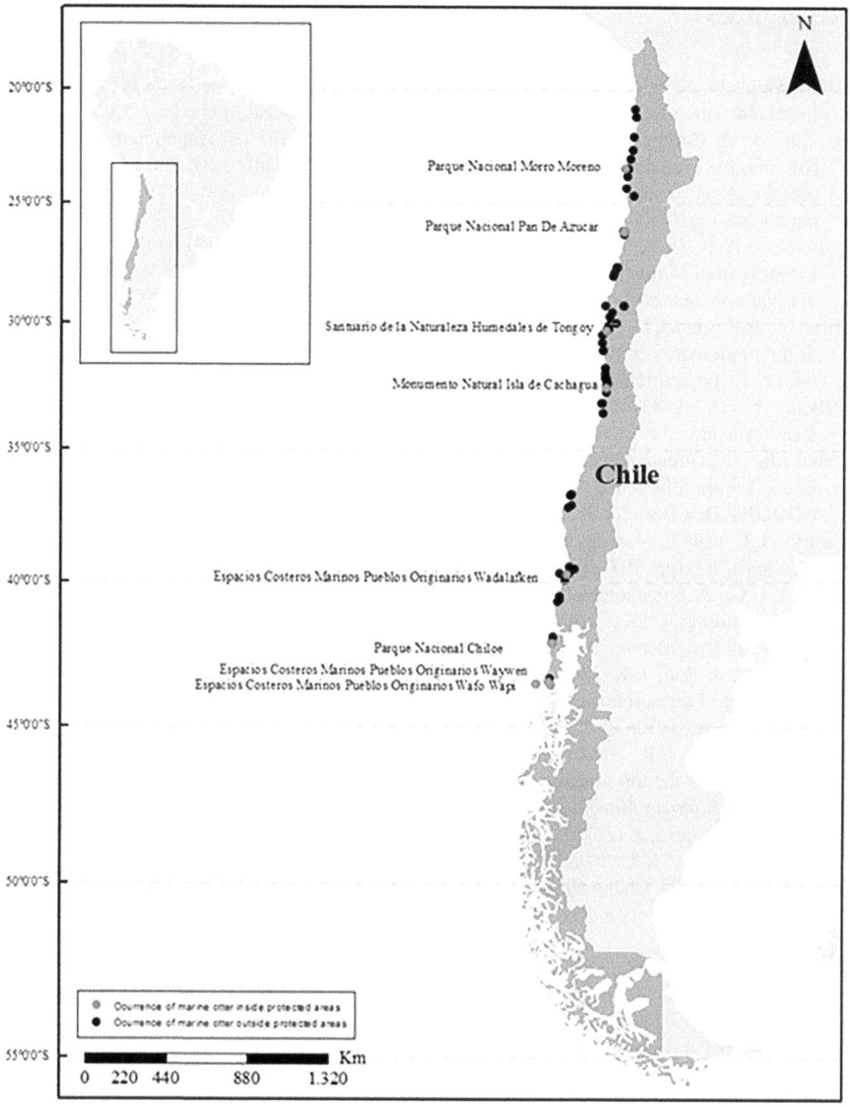

Fig. 8.4 Occurrence of the marine otter *Lontra felina* with respect to protected areas (SNASPE) in Chile. The black dots correspond to the points of occurrence of the marine otter outside protected areas. The green dots correspond to the points of occurrence of the marine otter within protected areas, these are eight protected areas: Parque Nacional Morro Morena, Parque Nacional Pan de Azucar, Santuario de la Naturaleza Humedales de Tongoy, Monumento Natural Isla de Cachagua, Espacios Costeros Marinos Pueblos Originarios Wadalafken, Parque Nacional Chiloe, Espacios Costeros Marinos Pueblos Originarios Waywen y Espacios Costeros Marinos Pueblos Originarios Wafo

References

Alfaro-Shigueto, J., Valqui, J., & Mangel, J. C. (2011). Nuevo registro de la nutria marina Lontra felina (Molina, 1782) al norte de su distribución actual. *Ecologia Aplicada, 10*, 87–91.

Badilla, M., & George-Nascimento, M. (2009). Conducta diurna del chungungo Lontra felina (Molina, 1782) en dos localidades de la costa de Talcahuano, Chile: ¿efectos de la exposición al oleaje y de las actividades humanas? *Revista de Biología Marina y Oceanografía, 44*, 409–415. https://doi.org/10.4067/s0718-19572009000200014

Barros, M., Pons, D. J., Moreno, A., et al. (2022). Domestic dog and alien north American mink as reservoirs of infectious diseases in the endangered southern river otter. *Austral Journal of Veterinary Sciences, 54*, 65–75. https://doi.org/10.4067/S0719-81322022000200065

Biffi, D., & Williams, D. A. (2017). Use of non-invasive techniques to determine population size of the marine otter in two regions of Peru. *Mammalian Biology – Zeitschrift für Säugetierkd, 84*, 12–19. https://doi.org/10.1016/J.MAMBIO.2016.12.006

Cabello, C. C. (1985). *Informe final Proyecto 1409 WWF/IUCN nutria Marina, Chiloé, Chile*. Ancud.

Calvo-Mac, C., Gutleb, A. C., Contal, S., et al. (2020). Exposure to toxoplasma gondii in marine otters (Lontra felina) and domestic cats (Felis catus) in an arid environment in Chile. *Journal of Wildlife Diseases, 56*, 962–964. https://doi.org/10.7589/2019-10-269

Castilla, J. C. (1982). *Nuevas observaciones sobre conducta, ecología y densidad de Lutra felina (Molina 1782)* (pp. 197–206). Mustelidae) en Chile. Publicación Ocas Mus Nac Hist Nat.

Castilla, J. C., & Bahamondes, I. (1979). Observaciones conductuales y ecológicas sobre Lutra felina (Molina) 1782 (Carnivora: Mustelidae) en las zonas central y centro-norte de Chile. *Archive of Experimental Biology and Medicine, 12*, 119–132.

Córdova, O., & Rau, J. R. (2016). Interacción entre la pesca artesanal y el depredador de alto nivel trófico Lontra felina en Chile. *Revista de Biología Marina y Oceanografía, 51*, 621–627. https://doi.org/10.4067/S0718-19572016000300013

Córdova, O., Rau, J. R., Suazo, C. G., & Arriagada, A. (2009). Comparative study of the feeding ecology of the top predator Lontra felina (Molina, 1782)(carnivore: Mustelidae) in Chile. *Revista de Biología Marina y Oceanografía, 44*, 429–438.

Correa, R., & Pizarro, J. (2023). Mortality of *Lontra fe*lina (Molina, 1782) in Chile (2009-2022) based on reports. *IUCN Otter Specialist Group Bulletin, 40*(3), 145–151.

Crooks, J. A. (2002). Characterizing ecosystem-level consequences of biological invasions: The role of ecosystem engineers. *Oikos, 97*, 153–166. https://doi.org/10.1034/J.1600-0706.2002.970201.X

Cursach, J. A., Rau, J. R., Ther, F., et al. (2012). Sinantropía y conservación marina: El caso del chungungo lontra felina en el sur de Chile. *Revista de Biología Marina y Oceanografía, 47*, 593–597. https://doi.org/10.4067/S0718-19572012000300022

Darwin, C. (1889). *Viaje de un naturalista alrededor del Mundo*. Tomo II.

de Ferran, V., Figueiró, H. V., de Jesus, T. F., et al. (2022). Phylogenomics of the world's otters. *Current Biology, 32*, 3650–3658.e4. https://doi.org/10.1016/J.CUB.2022.06.036

Ebensperger, L. A., & Botto-Mahan, C. (1997). Use of habitat, size of prey, and food-niche relationships of two sympatric otters in southernmost Chile. *Journal of Mammalogy, 78*, 222–227. https://doi.org/10.2307/1382655

Ebensperger, L. A., & Castilla, J. C. (1992). Selección de hábitat en tierra por la nutria marina, Lutra felina, en Isla Pan de Azúcar, Chile. *Revista Chilena de Historia Natural, 65*, 429–434.

Estes, J. A. (1989). Adaptations for aquatic living by carnivores. In J. L. Gittleman (Ed.), *Carnivore behavior, ecology, and evolution* (pp. 242–282). Springer.

Gay, C. (1874). *Historia Física y Política de Chile*. Zoología Tomo I. Maulde y Renou.

Gutiérrez, L., Vargas, F., Pinto, P., et al. (2019). Impact of human activities in habitat use and activity patterns of the marine otter (Lontra felina) in Central Chile. *Latin American Journal of Aquatic Research, 47*, 122–128. https://doi.org/10.3856/VOL47-ISSUE1-FULLTEXT-13

Housse, R. (1953). *Animales Salvajes de Chile*. Universidad de Santiago.

Lariviere, S. (1998). Lontra felina. *Mammalian Biology, 871*(575), 1–5. https://doi.org/10.2307/3504377/2600421

Mangel, J. C., Whitty, T., Medina-Vogel, G., et al. (2011). Latitudinal variation in diet and patterns of human interaction in the marine otter. *Marine Mammal Science, 27*, 14–25. https://doi.org/10.1111/j.1748-7692.2010.00414.x

Mangel, J., Alfaro-Shigueto, J., Verdi, R., Rheingantz, M. L. (2022). *Lontra felina.*

Medina, G. (1993). Activity budget and social behaviour of marine otter (Lutra felina) in southern Chile. In *Proceedings VI. International Otter Colloquium* (pp. 62–64). Pietermaritzburg.

Medina, G. (1995). Feeding habits of the marine otter (Lutra felina) in southern Chile. *Proceedings of the International Otter Colloquium, 6*, 65–68.

Medina-Vogel, G. (2010). Ecology of emerging infectious diseases and wild species conservation. *Archivos de Medicina Veterinaria, 42*, 11–24. https://doi.org/10.4067/s0301-732x2010000100003

Medina-Vogel, G., Delgado Rodríguez, C., Alvarez, P. R. E., & Bartheld, V. J. L. (2004). Feeding ecology of the marine otter (Lutra felina) in a rocky seashore of the south of Chile. *Marine Mammal Science, 20*, 134–144. https://doi.org/10.1111/j.1748-7692.2004.tb01144.x

Medina-Vogel, G., Bartheld, J. L., Pacheco, R. A., & Rodríguez, C. D. (2006). Population assessment and habitat use by marine otter Lontra felina in southern Chile. *Wildlife Biology, 12*, 191–199. https://doi.org/10.2981/0909-6396(2006)12[191:PAAHUB]2.0.CO;2

Medina-Vogel, G., Boher, F., Flores, G., et al. (2007). Spacing behavior of marine otters (*Lontra felina*) in relation to land refuges and fishery waste in Central Chile. *Journal of Mammalogy, 88*, 487–494. https://doi.org/10.1644/06-MAMM-A081R1.1

Medina-Vogel G, Merino LO, Monsalve Alarcón R, de J (2008) Coastal-marine discontinuities, critical patch size and isolation: Implications for marine otter conservation. Animal Conservation 11:57–64. https://doi.org/10.1111/j.1469-1795.2007.00151.x.

Medina-Vogel, G., Calvo-Mac, C., Delgado-Parada, N., Molina-Maldonado, G., Johnson-Padilla, S., & Berland-Arias, P. (2023). Co-occurrence between salmon farming, alien American mink (*Neogale vison*), and endangered otters in Patagonia. *Aquatic Mammals, 49*(6), 561–568. https://doi.org/10.1578/AM.49.6.2023.561

Ministerio del Ambiente. (2021). *Inventario nacional de especies de Chile*. In: DS XX MMA 2021 (17mo Proceso RCE).

Molina, J. I. (1782). *Saggio sulla storia civile del Chili. Del signor abate Giovanni Ignazio Molina. nella stamperia di S.* Tommaso d'Aquino.

Moreno, C. A. (2001). Community patterns generated by human harvesting on Chilean shores: A review. *Aquatic Conservation: Marine and Freshwater Ecosystems, 11*, 19–30. https://doi.org/10.1002/aqc.430

Moreno, C. A., Sutherland, J. P., & Jara, H. F. (1984). Man as a predator in the intertidal zone of southern Chile. *Oikos, 42*, 155. https://doi.org/10.2307/3544787

Ostfeld, R., Ebensperger, L., Klosterman, L., & Castilla, J. C. (1989). Foraging, activity budget, and social behavior of the south American marine otter, Lutra felina (Molina, 1782). *National Geographic Research, 5*, 422–438.

Pizarro-Neyra, J. (2008). *Mortality of the marine otter (Lontra felina) in southern Peru* (p. 25). IUCN Otter Specialist Group Bulletin.

Poblete, A. A., Górski, K., Moscoso, J., et al. (2019). Estimation of the diet of chungungo Lontra felina (Molina, 1782) in two locations of the coast of the Biobío region, Chile. *Gayana (Concepción), 83*, 1–9. https://doi.org/10.4067/s0717-65382019000100001

Rozzi, R., & Torres-Mura, J. C. (1990). Observaciones del Chungungo (Lutra felina) al sur de la Isla Grande de Chiloé: antecedentes para su conservación. *Medio Ambient, 11*, 24–28.

Santillán, L., & Caro, K. (2007). Mamíferos marinos en el Puerto de Salaverry y Zonas Adyacentes. In *Libro de resúmenes*. CONCIMAR.

Santillán, L., Saldaña-Serrano, M., & De-La-Torre, G. E. (2020). First record of microplastics in the endangered marine otter (Lontra felina). *Mastozoología Neotropical, 27*, 211–215. https://doi.org/10.31687/saremMN.20.27.1.0.12

Sielfeld, W. (1983). *Mamíferos marinos de Chile*. Ediciones de la Universidad de Chile.

Sielfeld, W. H. (1989). Sobreposicion de nicho y patrones de distribucion de Lutra felina y L. provocax (Mustelidae, Carnivora) en el medio marino de Sudamerica Austral. *An Del Mus Hist Nat Valparaiso, 20*, 103–108.

Sielfeld, W. (1990a). Características del hábitat de Lutra felina (Molina) y L.provocax Thomas (Carnivora, Mustelidae) en Fuego-Patagonia. *Investig en Cienc y Tecnol Ser Ciencias del Mar, 1*, 30–36.

Sielfeld, W. K. (1990b). Dieta del Chungungo (Lutra felina (Molina, 1782) Mustelidae, Carnivora) en Chile austral. *Investig en Cienc y Tecnol Ser Ciencias del Mar, 1*, 23–29.

Sielfeld, W. K. (1992). Abundancias relativas de Lutra felina (Molina, 1782) y L. provocax (Thomas, 1908) en el litoral de Chile austral. *Investig en Cienc y Tecnol Ser Ciencias del Mar, 2*, 3–11.

Sielfeld, W., & Castilla, J. (1999). Estado de conservación y conocimiento de las nutrias en Chile scientific. *Estud Ocean, 18*, 69–79.

Torres, D., Aguayo-Lobo, A., & Acevedo, J. (2002). Los Mamíferos Marinos De Chile: Ii. Carnívora. *Ser Cient Ina, 25*, 25–103.

Valqui, J. (2004). Comportamiento de la nutria marina Lontra felina (Molina, 1782) en una ambiente antrópico, la bahía de Pucusana -Lima, Peru. Universidad Nacional Agraria La Molina (UNALM).

Valqui, J. (2012). The marine otter Lontra felina (Molina, 1782): A review of its present status and implications for future conservation. *Mammalian Biology, 77*, 75–83. https://doi.org/10.1016/j.mambio.2011.08.004

Valqui, J., Hartl, G. B., & Zachos, F. E. (2010). Non-invasive genetic analysis reveals high levels of mtDNA variability in the endangered south-American marine otter (Lontra felina). *Conservation Genetics, 11*, 2067–2072. https://doi.org/10.1007/S10592-010-0069-9/FIGURES/2

Vianna, J. A., Ayerdi, P., Medina-Vogel, G., et al. (2010). Phylogeography of the marine otter (Lontra felina): Historical and contemporary factors determining its distribution. *The Journal of Heredity, 101*, 676–689. https://doi.org/10.1093/jhered/esq088

Villegas, M. J., Aron, A., & Ebensperger, L. A. (2007). The influence of wave exposure on the foraging activity of marine otter, Lontra felina (Molina, 1782) (Carnivora: Mustelidae) in northern Chile. *Journal of Ethology, 25*, 281–286. https://doi.org/10.1007/s10164-006-0024-x

Index

L. Ayala et al. (eds.), *Marine Otter Conservation*,
https://doi.org/10.1007/978-3-031-53931-2

R

S

T

MIX
Papier aus verantwortungsvollen Quellen
Paper from responsible sources
FSC® C105338

If you have any concerns about our products,
you can contact us on
ProductSafety@springernature.com

In case Publisher is established outside the EU,
the EU authorized representative is:
Springer Nature Customer Service Center GmbH
Europaplatz 3, 69115 Heidelberg, Germany

Printed by Libri Plureos GmbH
in Hamburg, Germany